# SELF-CONDITIONING
## — • AND • —
# SEXUALITY

# SELF-CONDITIONING

## — • AND • —

# SEXUALITY

## Practical Information about Sex and Sexuality

New Concept of Conditioning, Sanity Elements,
Belief System and E-spot

## DR. TAHA

**To order additional copies of this book, contact:**
Xlibris Corporation
1-888-795-4274
www.Xlibris.com
Orders@Xlibris.com
47531

# CONTENTS

To my parents; my family; my teachers; and to all humanity

# Preface

Sex, self, senses, and sanity are essential parts of human life. Without sex there would be no self, without self there would be no senses, without senses there would be no sanity.

This book is a journey through the self in order to understand human sanity and sexuality. It is based on my new concept of conditioning, which has a major impact on human sanity and sexuality. A sane human can be conditioned to acquire positive and negative characteristics.

This book is mainly based on my personal clinical experience and philosophy, which may not match other medical books.

# Glossary

**AIDS:** Acquired immune deficiency syndrome

**ANS:** Autonomic nervous system: it is part of human nervous system, which consists of sympathetic and parasympathetic system that work involuntarily.

**HIV:** Human immune deficiency virus

**IQ:** Intelligence quotient: it is a score derived from standardized tests attempting to measure intelligence. People with low IQ scores are sometimes placed in special-needs education.

**Ref:** Reference

**STIs:** Sexually transmitted infections

**STDs:** Sexually transmitted diseases

**TV:** Television

**WHO:** World Health Organization

# Self

The term "self," "being," "I," or "me" refers to the body and soul of a person. Self starts with a single fertilized cell, which multiplies into a human being. Human self is unique and is not identical even in identical or conjoined twins (Ref: 1, 2). There is no diagnostic test that can accurately identify people's self, secrets, and desires; and people themselves have difficulty describing their *self*. However, there is a difference between sane and insane human or other mammals. Sane humans have advanced learning skills, which make them the most intellectual animals. Learning plays a major part in human lives. It *conditions* most of their inherited behavior according to their environment (family, culture, and media). Infants are born with different inherited behaviors (instincts). They have different thresholds for each instinct by learning their sanity elements and inherited behavior condition according to their environment. People brought up in Asia, Europe, Africa, India, China, or America are conditioned to their cultural values, traditions, customs, habits, and ways of communication, voice tone, accent, expression, and expectation.

Just like other mammals, humans are created with five senses, which enable them to receive life's events and to survive independently. Without the five senses, there would be no self-identity, feeling, or pain; and humans would not survive alone. In contrast, the presence of the five senses can subject humans to physical and mental pain. Most of humans' conflict starts with abuse of human senses (feeling) such as disrespecting, ignoring, hating, insulting, harassing, bullying, abusing—emotional, verbal, physical, and/or sexual abuse.

Considering the five senses, the self can be divided into inner and outer self. The inner self represents human inherited behavior without the five senses such as people's behavior during hunger and thirst, while the "outer" self represents human reaction to the events received through the five senses while awake. However, both inner and outer self are linked to the sanity elements (below). They include feeling, thinking, memory, learning, understanding, emotions, desire, belief, will, and behavior. These elements are linked together, and any factor affecting this link could affect the self.

Although the sane human was created to be the most intelligent animal, she or he is the most controversial creature. They have more extreme emotions and behaviors than other animals. Humans are able to love and to kill and to build and to destroy at the same time. They are the only creatures who made the history and contribute in science, literature, art, and humanity and build many civilizations. They have unlimited ambitions to build better technology and future; but they also destroy many civilizations and have a long history of abuse and wars. Humans have been harmful to themselves and to nature more than any other animals. They have evolved from a tribe into cultures or nations, but never united; and since the beginning of the history, there have been continuous wars and invasions between nations.

Human belief systems have a major impact on their behavior. Although most religions share the same moral and ethical values, every human has a unique "self" belief system and people belief systems condition differently. Regardless of people's religious beliefs or disbeliefs, acquired learning can condition human belief systems to commit extreme moral and/or immoral behaviors.

There are selfless (love other self) and selfish (hate other self) instincts in every human. The lack or the excess of love without discipline and/or the learning of hate at early childhood can condition children to commit immoral or criminal behavior. Despite current, advanced civilization, prisons are still housing people who committed different immoral or criminal behavior. All prisoners are born innocent but conditioned by family, cultural, and media values to commit immoral behavior. Metaphorically, they are unlucky compared to millions of people who commit different types of abusive antisocial or criminal behavior outside prisons.

In addition to selfishness and belief system, the "self" fear from embarrassment, humiliation, shame, loss, stigma, power abuse and from law punishment has a great impact on secretive human behavior. Fear can push humans to lie, cheat, deceive, and fake their emotions in order to achieve their selfish desires. Men may fake their emotions in order to attract women, and politicians may make false promises to win more votes in the election.

# Soul

The term "soul" usually refers to the living part of the self, and the spirit represents a supernatural energy, which cannot be perceived by human senses. Therefore, it is impossible for humans to understand the structure of the spirit or any substance that can't be perceived. The soul leaves the body at death, which results from the lack of oxygen, and breathing of oxygen may represent the link between the soul and the living body. With the exception of breathing, people look dead during sleep. They lose their conscious and their five senses. They become powerless and vulnerable to danger. By sensory stimulation, a sleeping man regains his consciousnesses and his perception. However, even when awake, people's perception is limited; and they can't perceive things beyond their senses or "behind their back." Racist people may smile in the face of a foreign man, but at the same time, they may stab his back with vicious allegations or with a knife. The foreign man may not perceive the "evil" act of the racist people. Equally, we can't rule out the evil or holy souls in the universe. The monotheists (Judaism, Christianity, and Islam) believe that the first human, Adam, was created by God from mud and become viable by a blow of the Holy Spirit. This means that humans are still sharing parts of Adam's genes and the Holy Spirit. However, many nonbelievers feel that it is difficult to accept facts without evidence.

During life, the soul sticks to human body and doesn't only represent the consciousness or the energy that activates the body but also represents human identity. After a prolonged sleep or deep coma, the soul activates the human body with the same identity and personality. Even during sleep and despite the lack of consciousness, the soul may struggle with bad dreams or nightmares by making twists and turns. Some people talk, scream, and/or develop symptoms such as sweating, rapid heart rate, and sudden arousal during nightmares. In addition, after puberty, men may develop an involuntary erection during sleep; and some of them develop ejaculation or wet dreams (nocturnal emission) within seconds and without any physical or sexual activity while the same men may struggle to reach orgasm while awake or conscious. Also, despite unconsciousness and lack of mental power during sleep, some men might be able to cease their nocturnal emission before it happens while anxious men who try to avoid nocturnal emission may develop wet dreams more often than nonanxious men. Nonetheless, there are different levels of sleep. Most people awake from sleep by sensory stimulation while general anaesthesia induced deep, painless sleep, and an overdose of anaesthesia may reach a fatal level during which the soul leaves the body. Metaphorically, after death, some souls may remain in the minds of the other people and may inspire "recycle" in their souls.

## Body

The body is the perceived part of the self and the vehicle for the soul. It consists of cells and fluids. The cells form organs, and the two main organs I would mention here are the heart and the mind.

## Heart (Love)

Over the time, the heart has been linked to love in poems and is regarded as the organ of love. Probably because it occupies the center of the chest, people feel their love emotions come from the chest. They tend to hug the people they love. Nonetheless, the heart is the most important muscle for human life. Without this pumping machine, there would be no pulse of life and no blood flow to feed every cell in the body with oxygen and nutrition. Also, heartbeats are closely influenced by human emotions such as love, fear, anger, happiness, and sexual excitement. There is no life without heart and love instinct. The latter is one of the inherited instincts in humans and animals. It exists even in the most vicious animals. The lioness shows love and care to her cubs. She feeds and protects them. She also plays with them. Animals from different species, such as cats and dogs, can love each other if they were brought up together. However, brothers and sisters who were brought up together in the same house may become enemies, abusing or killing each other. This is because love in the human is influenced by the conditioning of the sanity elements (below), especially belief system, feeling, and desire. Human conditioning is influenced by the environment factors (family, cultural, and media values). Although love instinct is closely linked to sexual desire, most people don't have sex with everyone they love. Humans are conditioned in early childhood to be heterosexual, to get married, and not to have sex with their family members even if they love them. However, if incest and homosexuality are acceptable in human culture, people would have sex with everyone they love. Conditioning of the sanity elements can enable a husband to kiss or abuse his wife. It can also enable him to fake his emotion of love.

Nonetheless, people are born with different love thresholds and are differently conditioned. Although love instinct threshold varies in each person, people can't live without love. Love plays a major part in children mental and physical development; and without love for life temptations (having family, moneys, friends, partners, power, position or fame), adults may become depressed and commit suicide. Children develop strong love to their parents, and to love deprivation in early childhood can make them detached from their family or society and develop multiple negative emotions such as fear, shyness, insecurity, anxiety, depression, hate, anger, paranoia, or distrust. It can also make them abusers or victims of abuse. In contrast, excessive love without moral discipline in early childhood may condition children to become dependent, selfish, careless, moody, impulsive, destructive, and/or abusive. The conditioning of love is more extreme in people living in cities compared to people living in the jungle, and it relates conversely with selfishness. Selfish rich, urban people usually hate to donate part of their fortune to the poor people.

## Mind (Nervous System)

The nervous system consists of the following: central nervous system and peripheral nervous system.

### Central Nervous System

Central nervous system (CNS) includes the brain, which is protected by the skull; and the spinal cord, which is protected by the vertebrae. Both the brain and the spinal cord consist of white matter, which contains bundles of very thin fibres (axons), and grey matter, which contains masses of the nerve cells (neurons). The nerve fibres connect with nerve fibres from other nerve cells by joints (synapses), forming a massive network of fibres, which serve as electric wires transmitting neurological impulses and signals between the brain and the rest of the body when awake and during sleep.

As I mentioned above, the brain can't respond to any external sensory stimuli without the five senses, and humans can't feel or differentiate the pain induced by a needle or by a knife without skin sensation. However, pain in human is a sensory feeling and cannot only develop by the skin cut or injury, but it can also develop by receiving uncomfortable sensory stimuli through the other 4 senses. Watching and/or hearing bad news, insults, threats, and racial, emotional, or physical abuse can induce painful, long-lasting memory and can provoke negative emotions and/or psychosomatic symptoms. Equally, pain induced by sexual abuse can't be healed by a daily dressing or painkillers.

## Peripheral Nervous System

Peripheral nervous system (PNS) consists of multiple long nerves driven from the spinal cord to serve the limbs, skin, internal organs, and genitals. There are thirty-one pairs of spinal nerves: eight cervical, twelve thoracic, five lumbar, five sacral, and one coccygeal. PNS is subdivided into two systems: the somatic and the autonomic.

### Somatic Nervous System
The somatic nervous system (SNS) is responsible for controlling the body movements that are under conscious control such as moving our skeletal muscles. It consists of neurons linking the CNS with the muscles, skin, and sense organs and keeps the body in touch with its surroundings

### Autonomic Nervous System
The autonomic nervous system (ANS) regulates body activities under unconscious control. It links the CNS with the viscera and the endocrine glands and maintains homeostasis in the body. The ANS reacts rapidly during exercise or extreme emotions of love, anger, and fear. It also reacts unconsciously within seconds during sexual desire. The ANS is subdivided into two main systems: the sympatheticparasympatheticsystems.

## Sympathetic

The sympathetic nervous system consists of a longitudinal nerve plexus located on the side of the spinal cord. It responds within seconds to fear or danger by "fight, flight, or faint." Unlike animals, humans may faint during extreme emotions, and fainting can occur as result of conditioning of their sanity elements. In acute fear, the sympathetic nervous system causes the following changes: secretion of adrenaline, acceleration of heart rate, dilatation of the pupils, dilatation of the bronchi, inhibition of saliva secretion, inhibition of digestive secretion and peristalsis, conversion of glycogen to glucose, and inhibition of bladder contraction. The sympathetic system also gives the sense of excitement one feels due to the increase of adrenaline in the blood and is responsible for sexual orgasm in both sexes.

## Parasympathetic

The parasympathetic nervous system includes the tenth cranial nerve (vagus) and the splanchnic nerve. The former receives sensory information from the chest and the abdomen, while the latter originates from the sacral region of the spinal cord and supplies the genitals. The activities of parasympathetic nervous system are opposite the sympathetic activities and include slowing of the heart rate, constriction of the pupil, constriction of the bronchi, stimulation of saliva flow, stimulation of the digestive secretion and peristalsis, contraction of the bladder, dilatation of blood vessels, sexual arousal in both sex, and penile erection in men.

Although they work in opposite directions, the parasympathetic and sympathetic nervous systems provide a complementary action that regulates unconscious human activities. For example, during acute fear, the sympathetic system starts first by accelerating heart rate, but the progressive increase in the heart rate for few hours may induce heart failure. However, the parasympathetic system takes over the sympathetic action and slows the heart rate back to normal. Similarly, during sexual arousal, the parasympathetic causes increased heart rate, blood pressure, enlargement of sex organs while the sympathetic induces orgasm (and ejaculation in men) and subsequent relaxation and stabilizes the heart rate and blood pressure.

Sexual desire in humans starts after puberty when the sex hormones increase and the skin of the genitals becomes sensitive to touch, and an ANS reflex develop, causing sexual arousal in both sexes and voluntary and involuntary penile erection and nocturnal emission (wet dreams) in men. However, sexuality and sexual orientation are influenced by the competency of the five senses (especially vision and hearing) and by the conditioning of the human sanity elements (especially belief systems, feelings, and emotions) during early childhood. Children are born with different interests and physical traits. The nondisabled children develop different physical and emotional attractions compared with blind and deaf children. The latter would have no exposure to visual sexual sources and may acquire limited sexual experience compared with the nondisabled children.

## Endocrine Glands

The term "endocrine" means ductless or glands with no ducts; hence, these glands secrete their product directly into the blood stream. In contrast, exocrine—such as salivary glands, sweat glands, and glands within the gastrointestinal tract—have ducts. The endocrine glands secrete substance called hormones, which regulate human body metabolism, tissue function, and body growth. Together with the CNS and ANS, the endocrine glands play an important role in human emotions and desires.

The main endocrine glands related to sexual functions are the following:

1. Hypothalamus is located at the base of the brain and produces hormones that control other endocrine glands such as thyroid gland (thyrotropin-releasing hormone), adrenal glands (corticotropin-releasing hormone), and the gonads (gonadotropin-releasing hormone).
2. Pineal is a small gland at the base of the brain, and it produces melatonin.
3. Pituitary is also located at the base of the brain, but anterior to the pineal body. Pituitary gland has three parts or lobes:

   • Anterior lobe, which secretes the following hormones: growth, prolactin hormone, adrenocorticotropic, thyroid-stimulating, follicle-stimulating, and luteinizing
   • Posterior lobe, which secrets the following hormones: oxytocin, vasopressin (antidiuretic hormone)
   • Intermediate lobe, which produces melanocyte-stimulating

4. Thyroid gland is located in the anterior part of neck and produces the following hormones: thyroxine or tetraiodothyronine (T4), triiodothyronine (T3), and calcitonin.
5. Pancreas is located in the upper abdomen and produces insulin, glucagon, somatostatin, and pancreatic.
6. Two adrenal. Each gland is located above each kidney and has two parts:

   • Outer part (cortex); produces glucocorticoids (cortisol), mineralo-corticoids (chiefly aldosterone), and androgens, including dehydroepiandrosterone (DHEA), precursor of estrogen in female, and testosterone in male

- Inner part (medulla); produces adrenaline (epinephrine), noradrenaline (norepinephrine), dopamine, and enkephalin.

7. Men have two testes, and they produce the male hormone (testosterone).
8. Women have two ovaries, and they produce the female hormone (estrogen).

Some of the endocrine glands are linked and signal each other in sequence, and this is called axis. For example, there is a link between the hypothalamic, pituitary and the adrenal glands called hypothalamic-pituitary-adrenal axis. Such axis can be influenced by human emotions such as depression. The latter can cause decrease in appetite and sexual arousal in both sexes and can stop the menstrual period in women. In addition, diseases or tumors of the endocrine glands can affect human sexual performance in both sexes such as diabetes, hypothyroidism, and Addison's disease.

## Senses

The five senses (vision, hearing, touch, smell, and taste) are the sensory receptors that enable human to "feel" life. We receive all life's events through our senses; and without these senses, humans would have no perceptions, pain, life memories, social identities, dreams, goals, sexualities, desires, or taste of life. The five senses have an important role in filling our memory store with lifelong learning experience. They are the main source of human feeling, and we can't feel life without these senses. They are also essential tools for human independence and protection from danger. Without smell and taste, humans can't differentiate clear water from corrosive acids or gasoline; without touch, humans can't feel the difference between a needle and a knife injury; and without vision, touch, and hearing, there would be no communication, judgment, wars or civilizations. The presence of the five senses can be the source love and hate, pleasure and pain, peace and war. Most human conflicts start with the abuse of people's senses "feeling," especially through vision and hearing. Disrespect, insult, and verbal abuse can inflict pain, anxiety, fear, avoidance, isolation, depression, insomnia, nightmares, hate, anger, counterabuse, or violence.

### Vision

Vision is a gift of light. Humans are blind without light; and with light, vision becomes the quickest tool for human learning, search, safety, pleasure, discomfort, perception, physical and sexual attraction. Vision is also the most powerful tool for proof, disproof, and discrimination. It is a continuous photographic receptor for life's events while we are awake. Even during sleep, people tend to dream of familiar visual images or events, while individuals who are blind since birth have no visual imagery in their dreams; but they may experience taste, smell, and touch sensations in their dreams. Vision is an essential tool for independence, rapid mobility, and safety. Without vision, it would be difficult to walk, run, swim, clean, paint, cook, build, invent, play sport, games, dance, ride a bicycle, drive a car or airplane, and be employed in many businesses. Almost all human judgment is preformed by vision. Through vision, one can easily judge if a person is coming to greet or kill. It can also easily identify people's age, gender, race, behavior, facial expression, and body language better than other senses. Vision can identify and discriminate the details of forensic material in every angle and in every corner better than any camera. It is impossible to understand or appreciate life—beauty, color, art, painting, dimensions, designs, architectures—without vision; and there would be no human civilization, transportation, technology, computer, electricity, lights, international communication without vision. In addition, vision not only helps us to identify objects, tools, and equipment but also helps sane human or scientists to understand, think, and learn their structures in order build or create better tools and equipment. People with strong visual memory are able to remember written numbers and words for many years, and that can help them to become highly intelligent in certain scientific fields.

Babies are blind at birth until they start to use their "brand-new eyes" for the first time after birth. It may take weeks or months for a baby to understand and adapt the color, shapes, sizes, and dimensions of the figures in their environment. Blind children, however, learn to use other senses for learning and thinking. Vision is the main source for physical attraction and desires. A blind child can't be attracted or won't fantasize over pictures or images that they can't see; and they are unable to accurately describe pictures or nature's images such as trees, buildings, tools, equipment, birds, clouds, stars, planets while sighted children show attraction to certain colors, toys, and objects. Gradually sighted children build up their interests, desires, and/or greediness. As they grow older, sighted children learn (condition) the criteria of the physically attractive female and male from fairy tales, their families, and/or the media; and gradually they build up their dreams and fantasies and become emotionally attracted to certain types of people. After puberty, their emotional attraction evolves into physical and sexual attraction. Gradually learning (conditioning) makes sighted adolescents sexually aroused (conditioned ANS reflex) to people with certain physical criteria, and their vision may become the main tool for "love at first sight." Vision may also impair their school performance. Looking at a nice female lecturer may make sighted students focus on her beauty rather than her lecture while a blind student may not react in the same way. Blind children or adults become more focused on understanding and thinking, and they may develop strong insights and determination to compete with sighted people despite lacking the visual learning skills and references. Ironically, vision can blind the insights of sighted people. Watching sex images may provoke sexual desire in sighted adolescents, but not in the blind; hence, vision became a very successful commercial tool to promote the sex trade. Even in public places, photos of sex objects may be found in the shops and the streets to promote the sale of many commercial products and newspapers. Equally, vision may trigger people's sexual desire if a sexually attractive person sits beside them in a public place, and many men keep their vision on women in the streets. The latter may encourage women to improve their look in order to get more male attraction. Historically, women's eyes are regarded as symbols of beauty in sighted people or in poems; and many women tend to wear makeup in order to improve how their eyes look, even if this involves removing their spectacles. In contrast, eyes are regarded as a main part of personal identity, and people may wear dark spectacles to disguise their identity.

Apart from sexual desire, vision is the main tool for human curiosity, searching, appetite, selfishness, and greediness for eating, drinking, gambling, shopping, and hunting. Women love to spend hours in the supermarkets; and men may also spend hours on watching sport, games, or the Internet. Going into a supermarket, both men and women may make many stops when their eyes catch a nice food or object in the market. Vision plays a big role on conditioning the ANS reflex and sanity elements. Watching delicious food at infancy may not produce ANS stimulus, but when the children start to eat different foods, they learn (conditioned) the different types of delicious food such as sweet and develop salivation (conditioned ANS reflex) on watching sweet or delicious food. Later on, their food desire may push them to steal sweets secretly while blind children may have difficulty to commit the same behavior. Vision can also affect sighted people's emotions and feelings more rapidly than blind people. Children are born without chronic fear; but through punishment, abuse, and threats, they learn (conditioned) the source of fear and develop symptoms of acute fear (conditioned ANS reflex) toward certain fearful objects or events, and some of them may develop panic or phobia from certain objects or events. Vision transmits graphic images of abuse more rapidly than other senses, and children may be affected by watching domestic violence such as their father abusing their mother. The visual experience could create a painful long-lasting memory, which can later on affect their behavior, school performance, and sexuality.

Vision can rapidly trigger our negative and positive feelings, and it also helps us to understand other people's feelings and emotions. Watching powerful people killing innocent people in a war can provoke our anger. Similarly, watching a racist person or news may trigger negative feelings within seconds while meeting a nice person, watching a nice movie, or visiting nice places may trigger positive feelings.

Without vision, there would be no urge or compulsion for food, gambling, TV, media, and Internet addiction. Women would save many hours of shopping; and men would not be able to hunt women, birds, fish, and wild animals or assassinate their rivals. However, without vision, we would also live in eternal dark, couldn't appreciate the beauty of life; and there would be no civilization, or technology. Further, vision has been used as tool for magic and beliefs. Visual illusions and magic may deceive people's perception and change their beliefs. Miracles have been an important scriptural subject in many faiths, and ironically, the eye has been regarded as a tool for envy or curse in many ancient books and faiths.

## Hearing

Although hearing loss may sound not as traumatic as visual loss, it may put a human at risk of death. Unable to hear traffic noises or the fire alarm may lead to fatal accidents by motor vehicles or by fire, respectively. Hearing loss disables a person from hearing vicious people gossiping and planning to kill him or charge him with false allegations. Further, hearing loss at childhood may impair speech, learning, communication, and confidence. Deaf people may get frustrated to express their emotions such as love, romance, sympathy, hate, anger by sign language. They also are unable to hear birds singing, the sound of music, or the "rhythm of life." It might be also difficult for nondeaf people to call deaf people from a distance or to wake them from their sleep by calling their name. In contrast, hearing noises such as screaming, swearing, yelling, drilling, traffic, airplanes, siren, snoring, insult, humiliation, intimidation, and political propaganda that stir hate; and racism on a daily basis may provoke distress, anxiety, anger, or violence.

Sight and vision loss at birth would be more traumatic and disabling than losing only one sense. Blind and deaf people may only communicate by touch language, and they would be dependent and unable to look after themselves. They may also have limited physical and sexual attraction and activities compared with able people. Even if they get married, they would be unable to look after their children and hear their voices when they need help.

## Touch

People touch each other's skin mainly for the following purposes:

1. Shaking hands (greeting)
2. Express empathy (holding hands, hugging)
3. Expressing love (holding, hugging, kissing)
4. Sexual stimulation and intercourse
5. Physical and sexual abuse
6. Awake people from sleep by touching
7. Examination (medical, legal, accidental)

Without skin sensation, we cannot recognize if our back is hit by a needle or by a knife. Hugging is an expression of love and passion. Mother touches her newborn for many months, and her touch creates a trustful relationship with her baby, but this trust may disappear if the mother neglects her baby. Hugging and kissing can build the trust between children and their parents while physical abuse provokes their anger or distrust. Further, sexual abuse may cause very mild skin bruises, but lifelong mental damage. After puberty, the genital skin becomes sensitive to touch, and rubbing genitals can induce sexual arousal in both sexes.

## Taste

Tasting food after birth may not mean much for a newborn. However, after weaning from milk, babies start to taste new kinds of meals; and gradually infants start to develop interest to certain food, especially sweets. Their food desire may push them to eat sweets, chocolate without their mother's knowledge. Many children from broken families or those who were subjected to emotional, verbal, physical or sexual abuse may find temporary pleasure and comfort in food; and later on, they may develop eating disorder. Equally, some parents may spoil their children with food, and they may also develop eating disorder. Nonetheless, as food desire is associated with temporary pleasure, many people who feel bored, depressed, or lonely may become obsessed with thought of eating delicious food. Excessive eating, food obsession, and addiction to certain food might be an acceptable habit in many cultures; but obesity is undesirable in certain cultures; hence, some people may spend time, money, efforts on dieting, or become bulimic to maintain their desirable body weight.

## Smell

Smell senses can save life during a fire accident. Smelling the smoke may be the first warning sign of danger of fire or toxic gases. Smell sense has an impact on people's feeling, physical and sexual attraction. Smelling flowers or nice perfume may improve people's feeling while smelling dirt may deter them. Smelling helps us to differentiate between fresh and decomposed food; and without sense of taste and smell, we can't differentiate water from oil, gasoline, or urine.

In general, the five senses have a direct impact on the self and on human sanity elements (below). They represent the receptors for human feeling, the key for people's emotions, and the source for their happiness and conflicts. Any behavior that leads to abuse of these senses in other people can induce negative feelings, or negative emotions such fear, hate, anger while human behavior that leads to respect people's senses such as care and love can trigger positive feelings and emotions. Human feeling is linked to other sanity elements (below) and can be improved rapidly by offering respect, greetings, empathy, apology, compliments, gifts, hospitality, and care. People's feelings can also be improved by using these senses for self-pleasure such as watching and listening to nice news or nice media while their feeling may become worse by watching or listening to bad news or racist media.

## Sanity

The term "sane human" refers to a person who has healthy and sound mind and can make rational decisions or legal judgments within the culture norms. Sanity may also represent intellect as humans were created to be the most intellectual animals and have the ability to learn from birth to death. Although medical and psychological books may define human sanity differently, I would like to divide sanity into ten elements as follows:

1. Memory
2. Feeling
3. Understanding
4. Thinking
5. Learning
6. Emotions
7. Belief
8. Desire
9. Will
10. Behavior

These elements are inherited in each sane human, but they have different thresholds in each person; and they are conditioned in early childhood according to the family, cultural, and media values. The five senses, especially vision and hearing, have an important impact on the conditioning of human sanity elements, the ANS reflex, and the immune system. Sanity elements are linked together, and any factor that interferes with one of these elements may interfere with human sanity. Negative emotions—such as fear, disrespect, hate, anger, and abuse—have a negative impact on human sanity, self-confidence, and behavior; and it may trigger extreme behavior and psychosomatic disorders.

## Memory

Memory is the storage of life's events, experiences, and skills that can help and protect a human to survive on daily bases. Without memory, humans and animals are unable to remember a simple safety task that can protect them from accidents. Humans have a unique long-term memory, which could extend from childhood till death. Their memory is gradually filled in the first few years of life and influenced by the conditioning of other sanity elements. Children from the same family have different inherited interests and different thresholds for storing life's events. Their memory is gradually conditioned by the acquired learning into a unique memory store, which represents their conscious self-identity. Infants are born with an empty memory store or blank paper, and most of their memories are acquired through learning. After birth, their blank paper is starting to be filled with events received through their five senses. Gradually, children's memory helps them to acquire many information, skills, and behavior, which enable them to be independent. Therefore, parents have a major role in filling their children's blank paper with knowledge and moral values.

The competency of the five senses is important in the way we acquire our memory. Metaphorically, children with no five senses have no significant acquired memory while blind children would not acquire visual memory, and deaf children would learn from their visual memory. However, with support, many disabled children may be able to use their other senses to learn and build a rich memory and insight better than able children. Sighted people may have different types of visual perception mixed with positive and negative emotions, which could trigger happy and sad memories for many decades or till death. Acquired visual memories associated with negative emotions may interfere with rational thinking. For example, sighted people may meet a blind date, and their rapid visual judgment could affect their emotions and disable them from understanding the whole picture of opposite person. Similarly, during a car accident, people's emotion may disable them from taking the registration number of the cars involved in the accident.

People have different memory strength, and powerful memory can enhance people's intellect. Equally, a powerful memory could be a source of happiness or source of distress, obsession, and misery. Certain memory can last from early childhood till death, especially if the memory is associated with extreme emotions such as excitement, fear, anger, or abuse. Such emotions may induce biological chemicals, which help to fix the events in the memory store for a long time. Although time may relate inversely with memory, people tend to hold painful memory for many years or till death. Painful memories of abuse during childhood may last for many decades and may induce phobias, panic attacks, nightmares, and personality disorders or traits. In contrast, lack of memory or dementia can make people dependent and unable to look after themselves or to remember names, time, date, places, events, or simple safety skills. A demented man may urinate in the bed as he forgets where to go when his bladder is full.

Memory store represents self-identity, and each person tends to store specific type of memory. The differences are influenced by people's age, gender, sanity elements, interest, family, and cultural and media values. Time and age may fade away many events from our memory. There is also a difference in the way men and women store their memories. Asking an old married couple about events that had happened to them in the past, each one of them may express different memories despite the fact that they were together during the events. Sighted people may store more visual memory than vocal memory; and they may remember emotional events of love, hate, or sex more than written words or numbers while blind people store vocal events and focused on the quality of the touch and the tone of the sounds. Blind and deaf persons would only be able to store tactile memory.

Memory is the source for thinking, learning, belief, understanding, imagination, fantasy, behavior, and desire. Children learn (conditioned) thoughts and develop their belief systems from their environment and according to their interests and needs. Some of their needs may not be fulfilled, and this can influence the way they store life's events. The memory of emotional deprivation or abuse may push some children toward unacceptable or secretive behavior such as smoking, excessive eating, alcohol drinking, illicit drug addiction, and sexual promiscuity. Sexual memory is one of the most powerful memories. This may be because of the following reasons:

- Secret sexual behavior is usually associated with fear and excitation, and these emotions induce biological chemicals, which help to fix sexual events in the memory store.
- Repetition also helps to fix events in the memory store. Listening to the same song many times for few months make people repeating them from heart. Similarly, people tend to repeatedly think about their sexual events, which make them more fixed in the memory store.
- Interest and pleasure may attract people's attention. Football fans tend to remember the name of the players and the details of the game more than nonfootball fans. Similarly, people with interest in sex tend to remember the sexual events more vividly than routine daily events.

Forgetting a painful memory is difficult and may require building a new happy memory and conditioning of the human sanity elements in a different environment.

## Feeling

Feeling represents a conscious reaction of the human instincts to the sensory stimuli received by one or more of the five senses. Feelings in the sane humans are conditioned by learning during childhood according to family and cultural values. After birth, newborns have no choice but to receive sensory stimuli through their brand-new five senses. The new sensations such as light, noises, temperature, or pain represent an acute change in the environment of newborns from the dark womb environment into a bright life environment. Newborns feel and react spontaneously to the bright light, noises, and cold or hot temperature of life's environment. Without the five senses, newborns would not receive stimuli and would not develop "external" feeling to life; but they may develop "internal" feeling to their basic biological need such as hunger and thrust. Feeling is linked to other human sanity elements and conditioned by the learned experience during childhood. As children grow older, their feelings become less spontaneous.

Pain represents an uncomfortable feeling. It may develop as results of abuse of the five senses and not only by skin cut. Children are born with different threshold or response to pain, and after birth, they react differently to external pain such as trauma and to internal pain such as hunger. Their feeling of pain later on is conditioned by learning. Children subjected to physical abuse by their parents may feel distrust or fear from their parents. Their feeling may develop into an uncomfortable, painful experience, which later on conditions their sanity elements and induces conditioned ANS fear reflex or fear when they see their abusive parents. However, the conditioning process in each child is influenced by the inherited threshold of their instincts and by the experience they acquire. For example, parents may threaten their children with snakes when they commit mistakes. Some of the children may develop a feeling of severe fear or phobia to snakes, and their ANS produces many symptoms of fear every time they see a snake. They may also develop nightmares from snakes despite the loss of the five senses (feeling) during sleep. Other children may not develop similar phobia or nightmares.

Feeling affects our daily life and our future; and it can be influenced by age, gender, experience, culture, time, environment, or events. Negative feeling can be triggered by watching, receiving, or hearing abuse while positive feeling can be triggered by watching, hearing, or receiving a compliment, or happy news. Even the feeling of pain or depression may disappear in some people when they hear very happy news or win a fortune. Equally, people's positive feeling can improve our feeling and/or our future. During a job interview, people may choose a candidate who makes them feel comfortable, confident, or trustful; and in a blind date, people may chose a person who makes them feel attractive or desirable. Coming across nice people at work may make you feel happy while working with racist staff can push you to change your job, career, or county. Good feeling is important for any long-lasting relationship, and bad feeling can lead to an extreme behavior. A happy husband may reward his family with gifts and holidays while if his wife committed adultery, he may feel upset, angry, or kill his wife and/or commit suicide. In contrast, sane humans can deceive their feeling, and a husband can deceive his wife with his love; but his real love emotion is with another woman. Also, his wife can pretend a feeling of pain to seek his attention. Similarly, a prostitute may pretend that she is feeling sexually aroused in order to attract more clients; and jealous people at work may pretend they love you, but behind your back they plan to destroy your career. Nonetheless, people cannot hide their feeling during acute crisis or during extreme emotions such as fear and anger.

The five senses have important impact on human feelings and emotions, especially vision and hearing. Blind and deaf children would not react to the threat of gun fire because they can't see or hear it, and they my not develop needle phobia or feel fearful from needle injection as they have not seen or heard a terrified child while receiving the injection. However, sighted children may develop terrified experience from watching other people feeling pain, and their feeling could develop later on into panic or phobia, which can affect their personal and social lives. Conversely, comfortable stimulus received by our five senses such as watching nice environment, comedy, sport, or listing to nice music, can improve people's feelings and can also improve many psychosomatic symptoms such as stress, anxiety, and depression.

Immoral learning may condition people to "feel insecure" abusing their power to gain their selfish goals. Selfishness makes them insecure, and pushes them to be greedy, or abusive. Rich people may feel insecure and become obsessed with investing more money despite having many billion, and powerful people abuse their victims to feel secure and dominant. Abuse of power is common in many countries. In civilized nations, powerful employers may project their management failure on their employees and sack them when they feel insecure while in less civilized nations, dictators abuse the rights of their citizens in order to make them feel powerless and submissive. Similarly, a father or an older brother may feel insecure without abusing and dominating their family and make them feel vulnerable and/or inferior.

# Learning

Learning is the process of acquisition and building the memory store with information, knowledge, skills, values, traditions, beliefs, facts, and experiences. Learning is an instinct in human and animals, but it is more advanced in human and has a major impact on conditioning human instinctive behavior. Learning in humans is a lifelong process, which starts after birth when the mother teaches her baby the direction of the nipple and stimulates the baby to communicate with her by smiling, whispering, talking, or singing. Gradually, learning conditions the baby emotions and communication skills according to their family behavior. They learn how to drink, eat, or talk according to their family teaching. Therefore, learning in childhood is like the carving on the stone that may induce fixed long-lasting behavior. Babies who used to eat by spoon repeat the same behavior in each meal and may find difficulty to force them to eat by hand. Some learned skills remain permanent for life and can't be erased from memory such as reading, writing, talking, riding a bicycle, swimming, or driving; other leaned skills may be weakened or forgotten. Without learning, babies may grow relying on their instinctive behavior, which is similar to the behavior of primitive people living in remote jungles.

Nonetheless, children are born with different thresholds for learning. Their learning is influenced by their interests, the thresholds of other sanity elements, and by their environment. Without interest, children may not study their homework. Hence, motivation and support is important in children learning. Even genius children may not develop their talent if they are not offered opportunity for learning. Children from the same family may have different thinking, IQs, belief systems, interests, and understandings. These differences conditioned their learning skills differently. Each brother may perceive information and knowledge in a different way. People who are able to read, write, and/or understand different languages may have the ability to learn more than the illiterates; and academic learning enables people to understand, interpret, and think more objectively. Their belief system may become more fixable to absorb, and they learn more knowledge or analyze life's facts in a more structured or rational way.

Learning has great impact on conditioning moral and immoral human behavior. Humans were created with selfish instincts such as greediness and jealousy. Immoral learning can condition these selfish instincts to extreme abusive behavior. Parents may spoil their children and ignore their abusive behavior to others. Without discipline, some of them may learn different immoral or criminal methods to achieve their greediness. Similarly, some parents may teach their children to hate other race or other faith. This hate may build up a fixed belief in the children and condition them to abuse people from other race or faith. Such immoral learning conditions their behavior and pushes them to abuse other people's rights for their selfish interests. Immoral learning associated with selfish interest may be subjected to greediness and addiction. Financially and/or politically powerful people may become addicted to the abuse of their power in order to gain more money and power. They "learn" how to use patriotic or religious slogan to gain publicity or votes from their followers.

## Understanding

Understanding is the ability to listen, perceive, analyze, interpret, clarify, store, and remember the given information or skills. Understanding is a key issue in learning and intelligence. The more information, ideas, and facts we are able to understand, the more we become knowledgeable and independent. The ability to learn and understand information has helped humans to protect themselves from danger and to build up civilization. Understanding and thinking can widen our knowledge, insights, and skills that help us to develop, build, adapt, invent, imagine, or create new things. Humans have a unique long-term memory, which can store lifelong information and experiences.

However, people's power of understanding varies; and their understanding is influenced by their interests, experience, knowledge, literacy, cultural values, beliefs, and language barriers and on the threshold of their sanity elements. Illiterate people are unable to understand written information, and literate people are unable to understand information given in a different language. Each culture has a certain way of verbal and social expression and communication, which can cause a barrier in understanding people from other culture.

The competence of the five senses can affect our understanding. Without sign language, deaf people can't understand verbal information; and without touching, blind people are unable to understand shape and size of different objects or read. In contrast, optic illusion or magic can deceive our understanding ability, and visual attraction can impair the understanding of the sighted people. Blind students may focus their attention on the lecture while sighted students may waste long time on watching other things in the lecture room or make visual judgment on people's physical look without trying to understand the lecture. Similarly, during a blind date, a sighted couple may ignore to understand each other just because they didn't like each other's looks. Receiving multiplesensory stimuli may impair our understanding. Listening to loud noises of the radio and/or the television may impair our understanding when we try to read.

Understanding is linked to other sanity elements, such emotion, belief, desire, and thought. Severe emotions can impair our understanding. During stress, anger, fear, or receiving bad news, people tend to lose concentration and are unable to understand what they have been told. Understanding is also influenced by people's belief system. Fanatic people may fail or refuse to understand simple moral facts about other religions. Their fanatic beliefs may paralyze their objective way of thinking and understanding. In contrast, misunderstanding of scientists can lead to a human disaster such as inventing harmful drugs. Misunderstanding can also affect people's reactions. A husband may misunderstand his wife's behavior when she talk to another man; and people may misunderstand facial expression, body language, and the ways of dressing of people from other cultures and make rapid false judgments of them.

## Thinking

Thinking is a conscious mental activity reflects a major part of "self" identity. It is a unique process even in identical twins; hence human culture may never unite. Thinking involves mental searching, planning, hoping, obsession, retrieving memory, interpreting, filtering, anticipating, and judging information or the events in order to reach answers, goals, behavior, or solve problems or puzzles. Imagination, meditation, and fantasizing are specific type of thinking restricted to humans and can have positive and negative impact on their personal and sexual behavior. Thinking can be unconscious, spontaneous, or triggered by curiosity or by sensory stimuli. At a certain situation, place, and time, an idea, thought, may suddenly develop in a person's mind. It may stay and evolve into a notion, story, song, poem, or invention; or it may disappear quickly without remembering. Animals have limited instinctive thinking, but humans' thinking is linked to their sanity elements such as long-term memory, desires, and advanced learning experience, which can help them to solve complicated problems. Most of daily human behavior are automatic (conditioned) and doesn't require thinking, but people tend to think after receiving attractive or uncomfortable sensory stimulation or when they need answers to their problems. Their fantasy or problems can be triggered suddenly while watching an event, working, resting, or while reading or writing. Major problems or crises may cause involuntarily racing thoughts and obsession which can trigger stress, anxiety, fear, anger, insomnia, nightmares. It may disturb people's peace, disable their positive thinking and/or lead to violent behavior.

Thinking has been linked with feeling, belief, and behavior (Ref: 3); and I *think* thinking is influenced by other sanity elements, age, gender, IQ, time, place, events, culture, experience, and the potency of the five senses. Thinking is influenced by acquired factors. The family, culture, or the media can "brainwash" people's thinking. Children start to think when they develop their memory store and belief system in the first few years of life. After they master the language, their curiosity pushes them to ask many questions and think about the events that attract their attention. Their thinking process is usually spontaneous, but it is gradually conditioned by learning of knowledge and the experience into personal thinking. Nonetheless, every child is born with a unique power of thinking, and some of them inherit high IQ in one field or another. Without learning and education, their high IQ may fail to develop. Academic education can improve people's IQ and help them to think in a more rational way.

Although there are many types of thinking—such as objective, critical, creative—people usually have their own ways of thinking. Their thinking differs when they are alone, at rest, or when they are busy or engaged with other people or with other activities such as speaking, reading, writing, cooking, eating, or performing sex. Some people are fast thinkers, and others may develop their thoughts and ideas over days or months. Deep thinkers tend to be independent and/or creative while poor thinkers tend to be superficial and/or dependent. Both of them can be idealistic or pragmatic, and both many not able to read the mind (thoughts) of each other.

Any thinking associated with fear, anger, or pleasure can cause obsession, compulsion, or addiction to certain behavior, rituals, or habits. Law and religions associated with fear and punishment can deter people from immoral or criminal behavior; but bias law or unfair fanatic belief can create, anger, and/or obsession which may impair rational thinking and push people for violence. Similarly, traumatic experience of abuse or racism can impair people's thinking and induce obsession, conditioned ANS fear reflex, phobia, or panic attacks. Fear of shame or fear of hell after committing a sin or a crime may provoke obsessive thoughts, guilt feelings, and compulsions to commit homicide or suicide. Equally, fear caused by certain terminal illness or grief may disable people's thinking. However, after a while, most people who were given bad news of their terminal illness change their thinking and accept death. In contrast, a thinking that is associated with pleasure may also lead to obsession, compulsion, or addiction. Certain people become obsessed with thinking of delicious food, pleasure of sex, gambling, alcohol, smoking, illicit drugs, pornography, especially when they are alone, board, or stressed. Their obsession may gradually change to compulsion or addiction.

Thinking is unique process, and identical twins think differently. It can be influenced by age, gender, experience, and cultural values. Young people think about their desires and goals more than elderly. Men and women may think differently; and people from different cultures are conditioned to certain types of values, traditions, and behavior. They may make rapid thoughts and judgment on people from other cultures or people practicing different behavior such as women behaving in a masculine way or men behaving in feminine way. Equally, every nation has a certain way of political thinking. Racist nations usually condition their people to have one way of thinking, which includes disrespect, or to hate people from other race or faith. Equally, fanatic people with fixed belief system have one way of thinking that includes disrespect people from other religions or nonbelivers.

## Belief

Belief is a basic instinct in human and in animals, and they would not survive if they would trust their enemies or beasts. However, humans have different belief system. Their belief is linked to other sanity elements and has a major impact on their behavior. Every person has a unique "self" belief system, which represents the certainty or the trust of the "self" with other natural or supernatural things. Human belief in perceivable and unperceivable objects may include one or more of the following: disbelief, doubt, suspicion, paranoia, relative or flexible belief, trust, and fixed belief. Vision is a rapid tool for judging visible objects, but people may my fake their emotions and some of them have difficulty to believe in things that they can't perceive by their five senses. As most humans live in cultures, their belief system is conditioned in early childhood by their cultural values. Environmental conditioning can change people's belief system from flexible to fixed and vice versa. I would prefer to divide human belief into two main types: spiritual and personal. Both types are linked together and influenced by other sanity elements. They are also liable to obsession and compulsion as follows:

*Spiritual belief (faith)* usually includes belief in the origin or the creator of the universe or in the life after death. Many people acquire spiritual beliefs, and most faiths share similar ethical values such as respect, care, giving, and forgiving and aim to control human desires, selfishness, greediness, and immoral or criminal behavior. Sanity makes human the most intelligent animal, enabling him to build civilization and technology and also enabling him to practise extreme abusive or criminal behavior. Vision can trigger human desires and make him greedy for life's temptation. The more temptation people see, the more they want to try. In addition, humans have "selfish" instincts such as greediness, dominance, and hostility, which can be a source for abuse and wars. Humans are the only speaking animal that are able to verbally abuse or insult each other. They are also able to have and abuse their power. Power can protect humans, and it can also destroy them. Without moral discipline, physically powerful children can verbally and/or physically abuse powerless children; men can verbally, physically, or sexually abuse women; group of powerful people, employers, leaders, or politicians can abuse or kill powerless people; and powerful nations can abuse the human rights of powerless nations and evade national and international law. The law is not always accessible or affordable and can't always protract the victims or their families. The law can't teach children ethics, but parents can. Hence, the first child abuse starts with the parents when they fail to teach their children ethics and moral values or fail to build a trustful loving relationship with their children. Selfish, ignorant, or fanatic parents may condition their children to be selfish or to hate other people who belong to other race, faiths, or subfaiths. Equally, certain faiths, cultures, or nations believe that their faith, race, or ethnicity are superior; and/or other races, faiths, or ethnicities are inferior, ignorant, or sinners. According to their fixed belief, they may invade or kill people from other races or faiths to fulfil their belief or to enter heaven. Metaphorically, God forgave Adam, but humans may never forgive each other. Fanatic people may kill Adam's generation or God's creations to enter God's heaven. In addition, human belief system correlates with selfishness. Selfish rich people and leaders believe in giving their money to people who support or vote for them, and not to the poor people who need the money. Even when they become religious, selfish people usually focus on practising rituals that "put them in heaven" and ignore practising the ethical values with other people. Further, any faith-associated fear can subject humans to obsession and/or compulsion of certain habits or rituals. It can also impair humans' rational thinking (see "Thinking") and affect their personal lives.

Nonetheless, every person is born with a unique "self" belief system, and their beliefs conditioned differently in different cultures. The monotheists (Jews, Christians, and Muslims) represent less than half of the world population. They believe in one God who had also been conceived as being the source of all moral obligations, the greatest conceivable existence, and the creator of Adam, the first man, made from mud and become viable by a blow from Holy Spirit. The monotheists believe that heaven is the place after death for the people who comply with the moral holy commands and practise their rituals while hell is the end for the sinners who follow their desires or abuse innocent people. Nonetheless, not all people are spiritual, and not all people with spiritual beliefs worship the same God or have a similar fixed belief system. Some believers may worship fire, sun, animal, evil or a supernatural power, or myths. Also, over the time, many faiths including monotheist have been divided into subfaiths with different moral values and rituals; and it becomes difficult to rule a nation by one faith. Hence, many nations introduce a secular law that can protect the human rights regardless their faith, race, ethnicities, or lifestyles.

In addition to religion, any object leading to excitement or pleasure can lead to obsession and addiction. In the last few decades, the media has become the "new religion" or the daily "heroin" for people's excitement and conditioning. Urban people from all age groups become addicted to watch television or to surf the Internet. As the "heroin" causes addition, people tend to be bored from receiving the "same dose" of the media pleasure. Children feel bored from watching the same cartoon, and adults feel bored from watching the same movies. Human boredom and curiosity push them to look for more exciting movies, sex, games, or sport. Ethical and religious programs become boring subjects for young people, who are conditioned to pleasure and excitement. They always look for more exciting movies with more violence, abuse, or bad language. Without moral discipline, children growing up watching such movies may copy the abusive behavior they see in the media. As they grow older, they can be conditioned by the commercial and racist media. Most of the commercial media is based on achieving profit by provoking human selfish desires while the racist political or fanatic media is based on provoking hatred against other races, faiths, or nations. They may disrespect the feelings of millions of people belong to other faith or race; at the same time, they refuse to be criticized in their belief or their behavior. Even during humanitarian crises, they focus on their "photo opportunity" to show their "hypocritical" efforts and ignore the real humanitarian contribution of other nations. Both types of media can condition young people's belief system to be selfish, greedy, moody, insecure, or racist. Some of the media-conditioned young people may commit extreme violent behavior or they may *act like God*, kill anyone against their "selfish" belief system. They may also form national racist parties to kill innocent foreigners. Similarly, leaders and politicians with "selfish" belief system, such as Hitler, used advanced technology during wars in order to kill massive number of civilians, without guilt or remorse. They usually make their casualties accountable or martyrs, but not the lives of the innocent they killed.

*Personal belief* represents the *trust* of a human with his surroundings (humans, animals, nature, spirit). It includes people's sensory perception, judgment, expectation, and anticipation. Personal trust is an instinct, and the presence of two or more strange people can always raise the doubt of trusting the other person. Over time, humans have never united. Their "selfish" belief system, fear, desires, greediness, anger, hostility, and jealousy make them insecure, distrustful, and disunited. However, humans can't always practise selfish or abusive behavior as they would not trust each other to build a family or culture. Even abusive dictators need to build trust with their teams to survive. In contrast, people can always practise ethical or moral behavior as it would promote their trust, peace, safety, and security. Hence, much of human personal belief is linked by the people's moral or spiritual belief. Trust is an essential part in any relationship within a family or within a human culture; and without trust, humans would not be able to build civilizations or technology. Personal belief or trust is also an important part for children's mental and physical growth. It links to other sanity elements and correlates positively with respect, care, and love and negatively with disrespect, deception, and abuse. Infants are born with different belief thresholds, and their first belief or trust develops with their parents or their caregiver. Parents have a major impact on conditioning the belief system of their children. A trustful relationship between infants and their parents provides safety, security, emotional stability, and attachment between infants and their parents. Conversely, neglect and abuse can make infants detached from their parents, abusers, or victims of abuse. Emotional depravation, punishment, or abuse can push children to lie or cheat to avoid punishment. It can also weaken their trust with their parents and trigger their negative emotions, such as hate and anger, toward their parents.

As they grow older, infants' belief system is being influenced by their environment and starts to change from spontaneous into a personal one. Gradually, they become more independent and judgmental in building personal or close relationships. Their trust with other people becomes either fixed or relative. Even adult couples who live together for many decades may not trust each other 100 percent in every aspect. They may have fixed trust in their mutual love but have relative trust in their skills. Equally, an employer may have fixed belief in local employees and relative belief in foreign employees. Nonetheless, humans' belief is linked to other sanity elements and liable to obsession and compulsion. Certain people have fixed belief in curses or blessings and may become obsessed with certain numbers, astronomy, or superstitions that give them curses or blessings. Their obsession may develop into phobia or compulsion of certain rituals, which could disable their personal and social lives. In contrast, some people who have a strong belief in their astrologists, spiritual leaders, or doctors may be cured by a fake drug (placebo) given by their spiritual leader or doctor. However, people's belief systems may "condition" over the time. Before 1960s, black people and gay men suffered discrimination and abuse; but with repeated human rights campaigns, people's belief system has changed in many nations. Similarly, smoking was a desirable or glamorous habit in the past, but public health awareness has changed people's belief and lifestyle. Smoking has been banned in many public places, and people start to exercise more than before. The global awareness has also made the new generation prefer listening to music rather than listening to the boring political hypocrisy.

## Emotion

Emotions are group of instincts present in human and animals, but they can be conditioned in the sane human to an extreme limit. Human emotions are either positive or negative, such as love versus hate, happiness vs. sadness, trust vs. distrust, content vs. greediness, excitement vs. boredom, peace vs. fear, care vs. anger, bonding vs. rejection. Children are born with both negative and positive emotions, but with different thresholds, and are conditioned differently by their environmental factors. Positive emotions are the main source for human moral behavior, safety, protection, and progress. In contrast, negative emotions can have long-lasting harmful effects on children's mental growth, sanity elements, mood, ANS fear, and/or anger reflexes. They can push human toward immoral and/or criminal behavior. Nonetheless, human emotions are influenced by gender, age, culture, faith, time, experience, and place. Cultural values can condition men's and women's emotional expression differently. In certain cultures, men are allowed to show their anger in public places but not their love, tears, or weakness. Also, there are physiological differences in gender. Women are more tearful than men and show more attachment, care, and love to their children. In contrast, children's emotions are influenced by their parents', caregivers', or family's emotions. Lack of positive emotions can have significant impact on the development of the sanity elements in the children. Lack of love, trust, care, and support may cause emotional depravation; and this can make children detached from their family, vulnerable to abuse, or abusive to other. It can also affect their mental development, undermine their confidence, and impair their school performance. Emotional deprivation can lead to vulnerability, shyness, fear, guilt, distrust, avoidance, stammer, tics, chronic anxiety, insomnia, panic attacks, phobias, mood swings, obsession, compulsion, depression, paranoia, antisocial behavior, sexual promiscuity, alcohol, and/or illicit drug abuse, aggression, violence, suicide, or homicide.

Unlike animals, humans may stay for days with negative or positive emotions, especially depression and anger. The former may lead to suicide, and the latter can lead to violence or homicide. Extreme anger can cause continuous alertness, distress, insomnia, and/or nightmares. It can impair human sanity elements such as thinking, understanding, learning, belief, and willpower. In contrast, extreme love can make humans submissive and vulnerable to their lovers' abuse. Spiritual belief can have direct impact on people's emotions and personal behavior. Fear from punishment or from "hell" can deter people from committing immoral behavior; but excessive fear can induce guilt feeling, anxiety, depression, obsession, and/or compulsion of certain rituals.

People can fake or deny their emotions for personal, financial, or political reasons. Hence, it may be difficult to trust strangers, salesmen, or politicians. Adolescent boys may fake their love emotions to win the heart of the girls. A woman may fake romance to a man to get his fortune. A political party may fake their promises before the election to win votes. A president or a prime minister may deny his sexual affair to avoid scandals. A prostitute may fake orgasm in order to please her clients. However, men are unable to fake orgasm and ejaculation without erection; and people are unable to fake their extreme emotions of love, happiness, fear, or anger.

## Desire

Desire is oneself's wish. Animals and humans have inherited desire for food and sex, but sanity can condition human desires to extreme limits. Desires in humans are also influenced by their interest, curiosity, boredom, and freedom and by the competency of their five senses. Without senses, humans would not develop attraction or desire. Vision has a major impact on physical and sexual attraction. It has a direct conditioning impact on sanity elements and can condition the ANS reflex toward certain delicious foods and certain desirable sexual objects or person(s). Seeing delicious food or erotic object can induce saliva and sexual arousal, respectively.

During infancy, children have instinctive desire and curiosity to touch and/or taste any object that attracts their vision. After weaning from milk, they start to taste different types of food and develop a desire to certain types of food such as sweet. Gradually, children conditioned by acquired learning develop different desires and goals. The threshold of their desires varies and is influenced by their inherited interests and by their acquired learning. Some of them develop strong determination to achieve all their desires or goals; others may fail to achieve any. In general, any desire associated with pleasure is liable to obsession and/or addiction; and without early childhood discipline, guidance, and support, vulnerable or spoiled children may become obsessed or addicted to desires associated with pleasure such as excessive eating. Similarly, vulnerable or spoiled adolescent may develop addiction to junk foods, smoking, alcohol, and/or illicit drugs. However, certain desires are more acceptable than others. Although excessive eating may lead to morbid obesity and ill health, it is an acceptable habit in many cultures. Excessive eating is also acceptable at work, but illicit dugs and sex are not. Also, food desire is more essential for human life than drugs or sex. People may die without food, but they may live a normal life as a celibate. Nonetheless, people's desires are influenced by age, gender, and culture and by the conditioning of the other sanity elements such as belief and willpower. Some of them may keep a bar of chocolate at home for months; others may not be able to keep the same bar in their shopping bag for a few seconds. Further, some people may develop desire for food, and at the same time they develop a desire to remain slim or sexually attractive. Hence, they may follow certain diet, practise exercise, or become bulimic (binge and vomit) to avoid obesity.

The contradiction of human selfish desires can have impact on their social lives and the lives of their children. Young people may have the desire to marry, but their desire may disappear after marriage by domestic rows; and they may split, regretting their desire. Equally, an infertile couple may develop a strong desire to have a child, and they may spend their fortune to achieve their desire; but after having children, they may get divorce and regret having children. In certain cultures, men have strong desire to have many children to work and support their families or to pay the expenses of their father's selfish desires of smoking or drinking alcohol. The abused children may grow up with strong desires to escape home, and this can make them more vulnerable to achieve their own desires.

Although sexual desire is not acceptable in public places, men are relatively more free than women to express their sexual desires. Men may make many stops while walking in the street to stare at desirable women. Their sexual desire may push them to visit sex shops or sex clubs. Some of them become bored from one sexual practise, and their curiosity and obsession can push them to try an extreme sexual practise or paraphilia. Their obsession and addiction can affect their domestic and social lives. Similarly, obsession or addiction to alcohol may push the alcoholic to drink on daily basis or in early morning in order to avoid withdrawal symptoms. Some of them may lose their family, children, relatives, friends, and work. They may become homeless, vulnerable to abuse and medical illness such as cirrhosis and cancer of the liver or of the gut.

Vision has direct impact on sexual desire. In the ancient time, strong men and beautiful women were desirable objects for sex, slavery, and fertility. In the last few decades, sexual images have become the main commercial source for promoting sex trade and prostitution. Television, satellite, and Internet have been used to stimulate sexual desire; and they can condition children and adolescents to different sexual behavior. With the accessibility of global media, sex becomes more accessible to young people and in many cultures. Boredom, loneliness, stress, and depression can push some adolescents and adults to watch pornography. Commercial sex media has also used scandals of politicians and celebrities to promote their trade. In contrast, many vulnerable young people develop conditioned desire to read the sexual affairs of their desirable celebrities and/or copy their sexual lifestyles.

## Will

Will is the self-power and the immunity to resist difficulties, crises, desires, and pleasure. It is the determination to achieve ambitions or goals. Willpower correlates inversely with selfish desires. Children are born with different interests and willpower. Their families and their culture have major impact on conditioning their desire and willpower. Moral discipline with support can build strong willpower and trust between the children and their parents while strict moral discipline, abuse, or lack of discipline can weaken children's willpower toward their own selfish desires. Moral values in most religions can strengthen willpower against selfishness, greediness, and life's temptations. Also, the trustful and supportive relationships between children and their parents can help children to develop immunity toward harmful desires. Similarly, parents' motivation to their children can enhance their children's determination and willpower. Some children develop a strong motive to compete with their peers. Their determination empowers them to resist many temptations and enables them to work hard in order to achieve their goals.

In contrast, spoiled children may develop boredom, short temper, tantrum, insecurity, emotional instability, selfishness, impulsive behavior, and poor willpower. Spoiled children may learn (condition) greediness to food or drink from their families and cultures; and some of them with may steal food, alcohol, or money to feed their habits and desires.

Willpower can be influenced by other sanity elements, especially belief and fear emotion. Fear from law punishment, cultural stigma, sin, and STIs may deter men from practising sex with desirable women. However, people are conditioned differently and have different urges for sex. Some may resist any sex temptation outside marriage; others may feel very weak toward certain sexual objects or person. Men with poor willpower for sex don't think about STIs risk, and they may commit rape to fulfil their sexual desire. In addition, people's willpower may impair during sexual arousal. Some of them might develop strong emotions during sexual foreplay that can impair their willpower and self-control and push them to have different types of sexual practise with many casual partners without using barrier methods or a condom. The latter is the main cause of STIs/AIDS pandemic. Further, alcohol and illicit drugs can have negative impact on willpower. Many cases of sexual assault and rapes had happened under the influence of alcohol and/or illicit drugs.

Poverty and emotional, physical, and sexual abuse may also impair people's willpower. Verbal abuse, humiliation, intimidation, bulling, marginalization, racism, power abuse, war, or invasion can impair people's willpower and push them to commit unacceptable behavior regardless their faith or backgrounds. Humiliation of a husband to his wife can push her to seek love and support from another man. People living in poor countries, who lack simple human rights and social support, may feel powerless to improve their situation or to survive. Poor children may be forced to work long hours with low payment. Others may find no work and may steal food or money. Poverty can push some people to sell their bodies or to kill for money. Equally, poor parents in poor countries may sell their children for adoption to rich people or to pedophilic tourists.

## Behavior

Behavior is the performance of the self, which determines human identity and intellect. It includes all the physiological and mental activities of a sane human when awake and during sleep. Behavior starts in the germ cells (sperm and ova), which make the embryo, till the death of the human being. In general, human behavior is conditioned, contagious, and reflective. Most of human behaviors are conditioned by his environment; even in the womb, the embryo may be affected by his mother's medical and environmental factors. And in order to simplify human behavior, I would divide it into two main elements:

## Inherited Behavior (Unlearned)

Each fertilized ova has a unique type of genes that gives the human specific inherited (instinctive) behavior. Instincts are present in all healthy humans and animals and can help them to survive. They are unlearned, automatic, irresistible, and triggered by specific stimulations. Some animals and insects have more superior inherited behavior than human such as migration in birds and collaborative work of bees. However, advanced learning skills in humans condition most of their inherited behavior according to their environment. Apart from the sanity elements, human share many instincts with high animals as follow:

1. Anger (reaction to sensory abuse), which may range from alertness, agitation, avoidance, isolation, hate, rage, violence
2. Attraction, physical and sexual (more in sighted person)
3. Beliefs which range from disbelief, doubt, suspicion, trust, fixed belief
4. Belonging (to a group)
5. Bonding (to a person)
6. Cleaning, grooming
7. Compulsion (e.g., for food, sex, revenge)
8. Curiosity (more in sighted person)
9. Desire (e.g., for food, sex, play)
10. Disgust (more in sighted person)
11. Expression (e.g., during happiness, sadness, anger, hostility)
12. Fear (reaction to sensory threat), which may range from surprise, startle, alertness, agitation, anxiety, escape, fight, fainting feeling, faint (fainting mainly in human)
13. Gazing (with or without thinking)
14. Hostility or aggression (more in male)
15. Interest (more in sighted person)
16. Learning (advanced in human)

17. Limbs movement (e.g., during scratching, touching, grasping, walking, fighting)
18. Love emotion; which may range from empathy, care, feeding, protection, hugging, kissing
19. Mood status, which may range from low, weeping, boredom, normal, high, laughing
20. Playing (more in children)
21. Producing sound (e.g., during danger or excitement)
22. Rejection, which may include neglect, desertion, abandonment
23. Searching (e.g., for food, sex, pleasure)
24. Seeking attention (more in sighted person)
25. Selfishness: pride, dominance, greediness, jealousy, blame, projection, hostility
26. Self-protection, which may include lying, denying, projecting, pretending, seducing
27. Sexual response (e.g., sexual arousal in both sexes, erection and ejaculation in males)
28. Submission (obeying commands)
29. Thinking (advanced in human)
30. Physiological reflex: body temperature reflex (e.g., warming on freezing, cooling at heat), respiratory reflex (e.g., sneezing, yawning), consciousness reflex (e.g., awake, alert, highly alert, startle, sleeping, dreaming, erection, and wet dreams in men), impulse (e.g., to eat, drink, breath, urinate, laugh, cry), nerve reflexes (e.g., twitching, scratching, limb jerk), suffering (e.g., after exercise or injury)

Learning skills and sanity elements in sane humans are superior to other animals. It includes many tools such as thinking, understanding, interpreting, anticipating, long-term memory, high IQ, fantasizing, imagining, inspiring, deep insight, creating, achieving, judging, motivating, writing, reading, expressing, speaking, and communicating with different languages including sign language; and human hands are able to achieve many written and manual works. Hence, humans are the most conditioning animals and able to learn and/or copy more behaviors than animals.

Although humans are born totally dependent and may die quickly without feeding or nursing, with learning, they can become the most powerful creature and be able to build or destroy civilizations. Learning can condition human's sanity elements into moral "holy" and immoral "evil" behavior. Higher animals have no ability to learn immoral behavior such as lying, cheating, deceiving, blackmailing, and betraying and are unable to kill for money or for pleasure. Teaching children immoral behavior such as hate, racism, and discrimination can condition them to abuse others or be victims of abuse. Immoral learning can also condition humans to commit an extreme criminal behavior. Equally, strict discipline, or learning without love and support, can lead to similar results. Punishment of the parents to their children can force them to lie or cheat to protect themselves. It can also break the trust between the children and their family and enable the children to *learn* or copy the abusive or immoral behavior of their parents.

Learning can also condition the human ANS reflex in fear, anger, and sex toward certain person or object(s). Newborns have no fear from wild animals, unless they *learn* about their danger. Children can be conditioned to fear of snakes or to love them (Ref: 4). Similarly, learning conditions the ANS reflex during anger and during sex. Although sexual arousal starts after puberty, conditioning at early childhood can influence human sexuality and sexual orientation. Certain instinctive behaviors are acceptable in children, but not in adults. Playing is an acceptable behavior in children, but not in adults. However, adults may transform playing into more acceptable behavior such as foreplay, wrestling, dancing, or sport.

Nonetheless, conditioning of the inherited behavior is influence by the environment. The following table shows the impacts of environment on people living a primitive life in remote, isolated jungles such as the Amazon and the behavior of urban people living in large cities.

| Conditioned Inherited Behavior | Remote Tribe | Urban People |
|---|---|---|
| Cleaning | Limited | Frequent |
| Defensive fighting | Manual | Weapon |
| Discrimination | Limited | Wide |
| Greediness | Limited | Wide |
| Selfishness | Limited | Wide |
| Bonding | Wide | Limited |
| Emotions (e.g., anxiety, stress, fear, anger) | Limited | Wide |
| Boredom | Limited | Wide |
| Child tantrum | Limited | Wide |

| Excitement | Limited | Wide |
|---|---|---|
| Psychosomatic disorders and obesity | Limited | Wide |
| Physical attraction to shopping and gambling | Limited | Wide |
| Sexual attraction to TV or Internet pornography | Nil | Wide |
| Crimes (e.g., theft, sex trade or trafficking) | Limited | Wide |
| Anxiety about bills, taxes, or mortgage | Nil | Wide |

People who live primitive life would have relatively minimal learning and conditioning of their inherited behavior compared with urban people. Even pets are conditioned in the cities. Domestic cats may ignore hunting mice when they *learn* that their masters will always feed them. Also, domestic cats and dogs can be conditioned to live together in harmony, and beasts can be conditioned to play with each other in the circus. However, human conditioning is more complex, especially in the cities. Urban children learn different types of acceptable, unacceptable, immoral, illegal, or criminal behavior; and gradually their sanity elements develop and are conditioned differently. They start to understand cultural values and the link between behavior and punishment or reputation. Besides the sanity elements, children are born with curiosity for searching, passion for playing, love, or emotional needs. Failure to fulfil children's needs may lead to emotional deprivation. The latter can also result from strict discipline or lack of discipline, which may make them bored, detached from their family, and/or curious to search, play, and share excitement somewhere else in their environment. Curiosity can push them to copy any adult behavior that gives them excitement or pleasure such as smoking or drinking alcohol. Gradually their sanity elements, including their feeling, are conditioned to their environment.

Unlike primitive people living in a jungle, urban people usually acquire a broad spectrum of positive and negative feelings such as pleasure, happiness, excitement, thrill, euphoria, attraction, arousal, embarrassment, shame, regret, guilt, grief, sadness, depression, hunger, thirst, feeling full, respect, disrespect, satisfied, dissatisfied, disappointment, hate, anger, rage, doubt, confusion, ambivalence, suspicions, fear, being terrified, panic, rejection, repulsion, disgust, jealousy, envy, pain, warmth, cold, nausea, weak, tired, sleepy, unwell, dizzy, feeling fainting, content, empathy, ramose, overwhelmed, surprised, stressed, annoyance, upset, superiority, aggressiveness, hostility, pride, inferiority, feeling powerful, feeling powerless or insecure. The diversity of human feelings can have direct effect on his personal and social behavior. A happy husband may reward his family with gifts, while during extreme anger, he may abuse them.

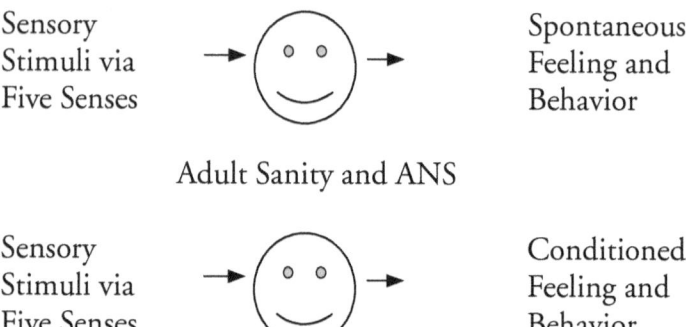

Newborn Sanity and ANS

Sensory                                         Spontaneous
Stimuli via                                     Feeling and
Five Senses                                     Behavior

Adult Sanity and ANS

Sensory                                         Conditioned
Stimuli via                                     Feeling and
Five Senses                                     Behavior

## Acquired Behavior (Conditioned)

There are two main elements for acquired behavior: cultural and global media. Both of them have a major influence on human behavior, including identical twins. Almost all identical twins have nonidentical behavior, and their behavior difference become more extreme if they are separated at birth and each twin is brought up in a different culture. Similarly, people living in India behave in different way from the people living in China, America, Africa, Europe, or Middle East.

## Cultural Behavior

Children's inherited behavior is conditioned by the sensory stimuli they receive in their environment. After birth, the infant's environment is dominated by the family first and then by the culture. The family have the major role in early children's conditioning. The conditioning is influenced by the competency of human senses, especially vision. Blind children would not be able to learn *visible* adult behavior or develop visual attraction like sighted children. They also develop different perception, interest, dreams, and desires. Curiosity can push sighted children to watch and learn their parents' behavior at home and adult behavior in public places from early age. Girls may wear their mother's makeup and high heels, and boys may copy their dad's behavior. They may smoke secretly to fulfil their curiosity. Gradually, the family or the caregiver conditions the inherited behavior of the sighted children according to their own values, beliefs, or wishes. Conditioning of children's inherited behavior can also be acquired by watching the same behavior in their environment for a long period of time. This may include the way their family behave when talking, eating, drinking, smiling, crying, yelling, fighting, loving, spitting, walking, and washing their bodies, their houses, or their streets. Hence, parents can sow the initial seeds of the personality, while human culture represents the land for the growth of personality seeds. Parents can spoil their children to be dependent, selfish, greedy, abusive, and criminals; or they can support them to be carers, creators, or heroes. Children who are born in happy, supportive families usually acquire happy moods while growing up in dysfunctional families may negatively affect children's moods and behavior. Equally, people who live in a friendly and sociable culture have less suicide rate while people who live in materialistic, competitive, or racist culture are usually less happy and may have a higher homicide and/or suicide rate.

As they start school, sighted children start to learn more behavior from their culture. Competitive urban cultures may induce competitive selfish personalities while rural cultures may induce less competitive personalities. Nonetheless, the prevalence of both behaviors depends on the proportion of the moral and immoral values present in their culture. Sighted children living in the cities tend to develop wide desire to see or try every thing that attracts their attention. Lack of discipline, boredom, abuse, or curiosity can push sighted children to try any behavior secretly or copy the behaviors that give them excitement and pleasure.

Urban human culture has been divided into two or more subclasses. Based on the economic status, people who live in big cities usually belong to at least three classes: common, middle, and aristocrats. Each class has different traditions, interests, hobbies, mentality, morbidity, and mortality. They also differ in the way they eat, think, and talk. They vary in their voice tone, quality of language, facial expression, and body language. The aristocrats usually frame themselves with "superior" behavior and may feel insecure to mix with the common. Common people who acquire money or prestigious position may change part of their conditioned behavior to adapt their new lifestyle and social position.

In general, most of human behavior is conditioned, contagious, and reflective. Children may copy their parents' or peers' behavior, parents may copy their culture's behavior, and employees may copy their bosses' behavior. However, nations' behavior can't easily be copied by other nations as each nation is influenced by their historical roots, traditions, customs, language, and political, cultural, and spiritual values. Human behavior is also contagious and reflective. Happy children are found in happy families and/or happy culture. Good behavior can bring good reaction feedback or compliments. In contrast, bad or selfish human behavior can cause distrust, fear, hate, anger, or violent behavior.

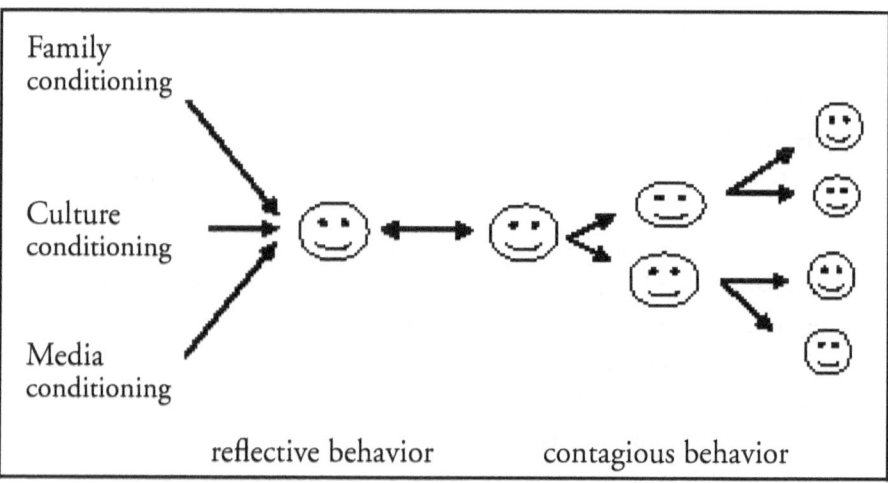

Human selfish instincts such as pride, dominance, hostility, and greediness are influenced mainly by vision; and sighted people living in the cities usually become more greedy for money, power, fame, fashion, and pleasure than people living in jungles. In addition, human behavior is unique and nonidentical even in identical twins. They may never agree. Historically humans have never united into one culture. Their sanity elements—specifically desires, greediness, anger, and fears—make them insecure, abusive, or victims of abuse. They may rob or kill an innocent person for money, and unlike animal, human greediness may continue after killing. The relatives of a rich person usually feel greedy in the fortune of their relative before and after his death. Moreover, although human immoral or illegal behavior can lead to social stigma or punishment, humans are able to commit any immoral or criminal behavior secretly or by "group power." The latter has been the most *acceptable* type of human rights abuse in civilized and less civilized nations. Groups of people, leaders, bosses, employers, politicians, professionals, organizations, or nations can lobby together and commit different types of unacceptable, immoral, abusive, or criminal behavior and evade the law. The power of the group has been the main source of human rights campaigns and human rights abuse. The lobbying of a few powerful politicians can determine the life of billion of people living in our planet. Greediness for power, fame, and/or money can push politicians to invade, abuse, or kill millions of innocent people in endless wars.

The diversity of human behavior has been ongoing since the beginning of history, and till now, the media disclose the scientific inventions and the criminal behavior of the humans on daily bases. Nonetheless, this diversity is influenced by age, gender, faith, disability, time, and cultural values. Human traditions change with the time and age, and personal behavior is more restricted in blind and disabled person while sighted people are more greedy to explore their selfish desires. There are also biological differences between men and women. Women are more passionate, tearful, and caring for children than men. Men are relatively less emotional and more hostile and determined than women in conducting their inventions and crimes, including homicide and suicide. The number of the male inventions and crimes are far more in men than in women, and certain criminal behavior such as rape and sexual abuse are mainly attributed to males. Men are relatively larger and are physically more powerful than women. This has enabled men to dominate women in hostility and in building civilizations, skyscrapers, dams, or tunnels. The physical power also enables men to commit more crimes, harassment, sexual abuse, and rape against women, children, and even animals. Ironically, powerful muscular male is more sexually attractive to most women. However, physical power does not explain men's aggressive sexual behavior. A physically weak man can still sexually abuse or rape a physically stronger woman by power of money or gun. The male hormone (testosterone) plays an important role in the male's sexual behavior. Castration of animals can inhibit their sex drive and aggressive behavior regardless of their size.

Besides the biological differences, the acquired factors have conditioned both men and women's social behavior. Human culture has set up certain behavior, traditions, body language, walk, dance, and dress code for men and women. Sex and nudity are not allowed in public, and human touch is restricted to hand shaking and hugging. Men are supposed to behave in a masculine way—be responsible at home, street, or work—while women are expected to behave in a feminine way. In many cultures, men behaving or dressing like women may get verbal or physical abuse. Also, men are not allowed to express their weakness or emotions while women may be allowed to cry or express their emotions. In addition, women are expected be attractive housewives and look after children and/or their husband. The latter has affected women's social behavior and makes them relatively insecure and dependent on men. Even in civilized cultures where women have more freedom, they may struggle between maintaining their independence, their femininity, and their personal relationship. Furthermore, vanity and waist size is not an issue in the jungle people; but the new civilization demands from men and women to be independent, physically and sexually attractive, and conditioned them to be more greedy and selfish for money and have high expectations. In contrast, moral conditioning can make urban people more sociable than jungle people. Love, care, and support can make children attach to their parents even after their parents' death. It can also make human friendships last forever.

## Global Media

The media represent a "sensory culture" for urban people. Humans' senses receive sensory (visual or auditory) stimuli of the media since early childhood. The media can have a major impact on human conditioning, but not so on blind and deaf people. The global media can condition human sanity elements, especially in sighted people, and change their perception and expectation. The conditioning impact of the media on young children can be more powerful than the conditioning power of their parents or their cultural values. The media has positive and negative impacts on human behaviors. It can inspire children to be professionals or criminals. In the last few decades, the accessibility of digital technology, cellular phone, and fast transport facilities has brought many instant benefits to urban people. It has also changed some of stereotypic human behavior and removed many cultural boundaries. However, despite the major advantages of digital technology in communication, scientific research, and inventions, it also has an ugly side. The television, satellites, and Internet have been used by the commercial and political media to sell their products and propaganda. The commercial media has become the new "heroin" that can give excitement and pleasure to millions of people every day, while the racist and/or the fanatic media can brainwash vulnerable children to become racist or abusers. Both types of media has become like cockroaches entering every home and bringing the "sensory pollution" with them. Different types of violent behaviors, bad language, racism, gambling, sex trade, and pornography have become available on the satellite channels and the Internet. Children at an early age can watch and copy adult behavior secretly. The pleasure of "heroin" and the excitement of the commercial media have made science, health, and moral programs less interesting. In addition, boredom, loneliness, and curiosity may push urban children and adults to look for new excitement in the media. Watching the same violent movie becomes boring, and people always look for more exciting movies with extreme violent behavior. Without family support of discipline, conditioned children may copy the bad behavior, bad language, or the sexual activity in the movies. Some of them may become curious about using illicit drugs or gambling. Addiction to such habits can push people to steal money or sell their body for money to feed their habits. In addition, the commercial media has changed people's expectations compared with the expectation of primitive people living in the jungle. Sighted people who live in the cities are more concerned about their superficial look and fashion. Some girls become victims of bulimia or anorexia. In contrast, some boys become obese or addicted to junk food. Equally, some adults become addicted to sex media or pornography. Furthermore, some people have been conditioned by the racist or fanatic media to hate or abuse people from other races or faiths. Racism can change people's feeling and behavior within seconds. It can push them to change the channel, or it can trigger many negative emotions and behavior such as hate, anger, and violence.

## Fear and Behavior

Fear is a reaction to a sensory threat and is a natural instinct in human and animals. Startle or sudden reaction to sensory stimuli can protect both humans and animals from dangers and helps them to fight or escape. However, fear instinct in humans is influenced by their sanity elements and can be conditioned into many chronic negative emotions; and unlike animals, extreme fear in human can lead to fainting. Infants are born with fear instincts, and they become fearful of any uncomfortable sensory stimuli, especially pain. Their fear threshold to pain varies, and its conditioning is influenced by environmental factors and by competency of their five senses. Blind and deaf children may not react to a fearful object they can't see or hear while sighted adult may become terrified from a tiny spider. Fear from punishment can trigger the ANS fear reflex in infant to induce involuntary symptoms of fear such as palpitation, tremor, insomnia, and/or nightmares. Frequent punishment can conditioned the ANS fear reflex to induce fear symptoms even before receiving the punishment. Conditioning of fear can affect children's behavior and personalities. Unlike animals, fear in human may cause stress, anxiety, shyness, distrust, avoidance, isolation, depression, guilt, hate, anger, insomnia, nightmares, phobias, panic attack, fainting feeling, and fainting. Fear in human can also induce obsession and compulsion.

Almost all humans' conflicts start with "sensory fear" or fear of abuse of the five senses "feeling." Power can protect human and can enable him to abuse others. Immoral powerful people usually disrespect, insult, or abuse the senses of powerless people. In contrast, without the fear from shame, stigma, embarrassment, failure, loss, power abuse, punishment, prison, law, or "hell," people would express their feelings, emotions, sexual desires, sins, or secrets to each other; and they can commit any criminal behavior including killing on a daily basis. However, even vicious criminals may develop fear while committing their crimes. The fear is linked to human sanity elements, especially belief, feeling, and emotions. Fear from punishment may push a child to lie; fear from getting fat may push a young girl to diet; fear from losing a job can make employees courteous; fear from divorce can push a husband to lie to his wife; fear from prison can push a thief to destroy the evidence of his crime.

Nonetheless, people are born with different fear threshold and are conditioned differently to the same types of fear. Although fear is more common in people with an anxious personality, nonanxious people may develop different types of fear and may react differently to pain. Pain is an unpleasant sensory feeling. People have different thresholds for pain, and their feeling of pain conditioned differently in early childhood. Not all people require analgesic after injury. Certain people fear from a mild skin scratch while others may not develop pain for many hours after a major traumatic accident (Ref: 5). In contrast, happiness or escape from a fearful event may relieve pain quickly. Wounded soldiers may not feel pain when they were transferred from the battlefield to the hospital. The joy of escaping death in the battlefield can ease their pain. Pain is unpleasant "sensory" stimuli and can be perceived by the five senses and not only by a somatic skin sensation. The "visual" and the "auditory" experience of death in the battlefield can trigger a painful "sensory" experience. Similarly, the sensory experience of watching children crying of pain while having a needle injection can induce fear of needle or needle phobia. The latter can lead to fainting in some adolescents or adults when they received a needle injection. Fear from authority, dictatorship, abusers, stigma, or shame can affect human's thinking, behavior, and performance and can also trigger psychosomatic disorders or illness.

Fear from law can help to promote peace and safety, but can't always control human immoral behavior. People can commit their crimes secretly or by lobbying with a group of people to achieve unethical or criminal behaviors. A group of staff can make false allegations to imprison their innocent colleague. Even a group of doctors are able to use their evil side to abuse their patients or their colleagues. Further, the law can't protect millions of women and their children from contracting HIV/STDs from men who hide their HIV-positive status and practice unsafe sex. Furthermore, the law is not always affordable or accessible. Seeking justice may require paying a high price and many months or years of waiting and mental suffering. The fear of discrimination and the stigma can deter abused people from reporting their abuse. They fear to lose their money, efforts, time, reputation, or position if they lose their legal case. Their fear can also trigger nightmares, depression, and many psychosomatic symptoms and/or may push them to commit suicide.

Fear can also affect human sexuality. Few adolescents who have a sexual experience with the same sex may develop severe guilt and fear of becoming homosexual. The fear can influence their belief systems and provoke doubts, chronic guilt, and obsession for many years. This can affect their personal and sexual lives and can conditions their sexual arousal (ANS reflex) only to the same sex. The latter can convert their doubtful belief system into a fixed belief system of being homosexual. Such conditioning may disable some adolescents or men from having an erection with a woman.

Fear can also condition people's sexual perception and can condition the feeling of pain to pleasure. Some people *enjoy* the "arousal" induced by fear, and the excitement results from the release of adrenaline during violent sex. Fear of pain may become a source for their sexual pleasure such as masochists, who usually enjoy violent sexual behavior with or without love. The thrill of adrenaline resulting from fear can also push many normal people to watch horror movies or to ride a roller coaster. Historically, in ancient Rome, people used to be entertained by watching gladiators killing slaves in the coliseum. In the last few decades, horror movies have made similar conditioning effects. Few people may become addicted to the excitement of abusing or inducing fear on others, or they may become sadistic.

Lack of fear can have negative impact on human behavior too. Children who grow up spoiled, without discipline or moral values may become dependent, greedy, moody, and impulsive with no goals or ambitions in life and have no appreciation of other people's achievements or contributions. They may acquire superficial values to fulfil their unstable mood. During childhood, they may show frequent tantrums; and during adolescence, they become insecure, impulsive, abusive, or self-destructive and may fake many emotions to seek other people's attentions. Their short temper can make them impulsive or violent to others when they experience fear or failure. Without moral discipline and family support, the pain of rejection or failure can push the spoiled people to abuse themselves (self-mutilation) or abuse other people.

selfish desires

↕

behavior  ↔  sanity elements  ↔  senses

↕

fear / lack of fear

*"Fear plays a major part in humans life especially when interferes with achieving their desires"*

# Personality

Personality is the overt part of the personal behavior. It represents the perceivable part of the self; hence, personality depends on people's perception and on the cultural norms. The family is the seeds for personality, and the culture is the land for the seeds to grow. Personality evolves and conditions in the family first. Although children are born with similar instincts, every child has different threshold for each instinct and is conditioned differently. Even identical twins who share similar genes develop different personalities (Ref: 2). As children grow older, culture will take over the family in conditioning people's personality. Children who are brought up in the cities are conditioned in a different way compared with those who live in the jungles. Children in the cities develop different goals, ambitions, and needs. The term "ego" has been misused to reflect many personal characteristics such as pride, superiority, or selfishness. Equally, the term "extrovert" and "introvert" have been used to reflect people's mood, but they can't accurately assess each personality. As people are conditioned differently in each culture, I would prefer to use human traits for assessing people's personality as in the following table:

| Fair | Unfair |
|------|--------|
| Merciful | Abuser |
| Unselfish | Selfish |
| Honest | Liar |
| Humble | Arrogant |
| Polite | Impolite |
| Generous | Stingy |
| Quiet | Talkative |
| Independent | Dependent |
| Deep | Superficial |
| Open-minded | Fanatic |
| Reliable | Unreliable |
| Patient | Impulsive |
| Benign | Hostile |
| Optimistic | Pessimistic |
| Clean | Dirty |
| Brave | Coward |
| Organized | Disorganized |

| Leader | Follower |
|---|---|
| Faithful | Unfaithful |
| Forgiving | Confrontational |
| Sociable | Loner |
| Clever | Stupid |
| Realistic, Pragmatic | Idealistic, Dreamer |
| Self-sufficient | Greedy |

From the above table, assessment of the personality can be achieved by estimating the proportion of each trait and summing the total of each side of the table. Negative traits such as being selfish, abusive, and/or greedy are products of civilization and, in certain cultures, can be regarded as signs of a strong personality. Nonetheless, people can fake their emotions and behavior, and their professional personalities at work may differ from their real personalities. A man can be a Mr. Perfect at work, but depressive and/or aggressive at home; hence, personality assessment may need months or years and can't always be accurately assessed on a blind date. Changing people's personality is a very difficult or impossible task and may require changing their belief system and their environment. Just like learning driving or swimming, certain behavior can't be easily erased from the memory, especially in selfish people with a fixed belief system.

With the globalization and the demographic changes of human culture, human's conditioning has changed in both genders. The increase in life's temptations has increased people's desires, obsession, compulsion, addiction, and sexual behavior. Human psychology has also changed, and new personality traits or disorders have evolved especially among young people such as a manic-aggressive-depressive disorder. A few young people who are brought up in competitive cultures may commit massive killing of innocent people then kill themselves. Similarly, fanatic people with fixed belief systems may be conditioned to do the same.

# Periods of Sexual Development

Human development or life cycle includes different physical and mental growth periods. It starts from the penetration of the sperm to the ova to form an embryo and till death. Although humans may live one hundred years, most of their physical growth occurs during the first two decades of life, while their mental development and sexuality are influenced by conditioning acquired in the first decade of life. Fear and love can have major impacts on human personal development and sexuality as follows:

## Embryo -0

After sexual or artificial fertilization, the fertilized ovum grows into an embryo. The inherited part of human personal and sexual behavior is present in the genes, which are located on the chromosomes of the fertilized ova. Genetically, each sperm has twenty-three chromosomes; and one of them is a sex chromosome, which is either type Y or type X. Similarly, a female ovum has twenty-three chromosomes but contains only type X sex chromosome. The Y chromosome is responsible for formation of the testes and its absence leads to the development of ovaries. Hence, in the uterus, the fertilized ova would either have male sex chromosome (XY) or female sex chromosomes (XX). If the Y chromosome is present, the gonads become testes; and they begin to produce androgen (precursor of testosterone), which stimulates development of the penis and the scrotum. Very rarely, a hermaphrodite or an abnormal number of sex chromosomes may present in one fertilized ova (e.g., OX, XXY, and XYY); and this can lead to different female-and-male morphology, ambiguous genitalia, or intersex. Also, the congenital abnormality of sex hormones in utero can produce ambiguous genitalia. Although these cases are rare, after puberty, they may develop sexual orientation problems and need long-term medical and psychological care and support. Fear, stress, and taking certain drugs or hormones during pregnancy can lead to hormonal imbalance in the embryo and may also affect its sexuality. However, after birth, environmental factors can play a major part in conditioning human sexuality.

## Infancy 0-2

This is the period of breast-feeding of milk and "sensory stimuli" or "feeling." After birth, infants begin to use their brand-new five senses for the first time after being in a dark womb. They start to perceive life's picture through their vision and adapt to their new environment just like solving an optical illusion or a jigsaw. Newborns have empty memory, and the sensory stimulation they receive by their parents or caregiver fills their memory store and conditions their immature sanity elements. Although babies are fully conscious in this period, most people don't remember the events that had happened to them in this early stage of life. This may be because their memory store has not developed enough yet or has not filled with enough information that can make interpretations to life's events. Hence, this period is crucial in filling the empty memory store with "comfortable" sensory stimuli that can build the trust or "self" belief system in the infants toward their parents or caregiver. Child neglect or abuse in this period can interfere with infant's mental development and build up fear and distrust between the children and their parents.

Fear is an inherited reaction to sensory threat, and after birth, newborns may *feel* terrified to face the sudden change in sensory stimuli of the "dark" warm womb and the "bright" life environment. They are powerless to change the environment and are totally dependent on their caregiver. Without feeding and nursing, they die within days. They may feel terrified from any uncomfortable sensory stimuli, and screaming is their main way of communication. Sucking the nipple makes them emotionally closer to their mothers, and frequent sucking makes their lips more powerful than their soft hands in grasping objects. Infants also feel and taste the food by their lips and tongue; and gradually their mouth becomes the most powerful sensory tool for grasping, feeling, tasting, screaming, and communicating. They usually use their mouth to test any foreign object they hold. At the end of this period, infants start to express different activities, attraction, and desires. They start to train their voice and limbs for expression and communication. Fear from abuse can create distrust, negative emotions, and delay in children's physical and mental development.

## Childhood 3-8

This is the period which starts after the initial conditioning of newborns by their family or caregiver. During this period, culture and the media also enter into the conditioning process of the children. Gradually, children become physically and mentally more independent—able to walk, talk, play, and express their feelings and thoughts. They use their five senses and their developing sanity elements to learn new skills and ask many questions to fulfil their curiosity about life's events. Gradually, they become *conscious* about the acceptable and unacceptable family and culture values. Hence, in this period, moral learning, and discipline, has a crucial role in conditioning children's instincts; and it sows the seeds of the human personalities. Like most animals, children are born with the instinct to play; and a sighted child develops different interests, activities, and physical attraction compared with a blind chid. Sighted children have more curiosity to copy adult's positive and negative behavior. They also seek their parents' attention and love and show curiosity to search their environment more than blind children. In a dysfunctional family, parents ignore children's needs, and this may lead to distrust and emotional depravation, which may make children detached from their family and vulnerable to strangers' attention and attraction. Conversely, parents may spoil their children to become moody, selfish, or impulsive. At school, children become more independent and start to make friendship, develop new interests, hobbies, wishes, and goals. As I mentioned above human behavior is conditioned, contagious, and reflective. Sighted children start to be conditioned by their family behaviors such as the way they talk, react, speak, argue, or laugh. They may also copy adult behavior that they watch in their environment.

Some of them may smoke and drink alcohol secretly or with their peers to satisfy their curiosity. Their poor relationships with their parents may push them to be closer to their peers, who enjoy playing with them and giving them attention or excitement. Although at this stage children cannot have sexual arousal or erection, it is not uncommon for them to play with their genitals and copy adult behavior. Such early learning experiences may influence and condition their future sexuality. However, first child abuse usually starts in the family. Vulnerable children are product of selfish, ignorant, or fanatic parents who are unable to look after children's emotional, educational, and financial needs. They may also abuse their children's minds with hate toward people from other races or faiths or subfaith. Emotional, physical, and/or sexual abuses are common in dysfunctional families. Fear from abuse can cause long-term negative emotions, distrust, withdrawal, shyness, phobias, avoidance, delayed speech, pica, night terrors, and incontinence and can also impair school performance.

## Puberty 9-17

Puberty and adolescence is the hormonal volcano period where sex hormones peak in both genders and cause physical, emotional, and behavior changes. It is a very critical period in human mental and sexual development. Puberty may begin in girls between the ages of nine to seventeen and usually precedes boy's puberty. It starts with changes in the brain or the hypothalamus (see "Endocrine"), which trigger the release of the hormones from the pituitary gland such as the growth hormone, which is responsible for physical body growth and gonadotropin hormones, which stimulate the adrenal glands and the gonads (testes in male and ovaries in female) to secrete estrogen, progesterone in girls, and androgens, particularly testosterone in boys. These hormones are responsible for development of sexual characteristics. The latter can be divided into two types: primary sexual characteristics, which are essential for reproduction such as sex organs or genitalia (penis and scrotum in male and clitoris and vulva in female) and secondary sexual characteristics, which include the characteristics that physically distinguish one sex from the other such as the larger breasts in girls and the facial hair and deeper voices in boys. A girl's breasts grow first, then body hair (underarm and pubic hair growth with increased secretions from oil and sweat glands). This is followed by the first menstrual period (menarche), and regular ovulation may start two years after menarche. Boys' sexual changes start after the girls with the increase in the size and the color of the scrotum, followed by appearance of pubic and underarm hair with increase in secretions from oil and sweat glands. This is followed by penile and facial hair growth and the deepening of the voice. Voluntary and involuntary penile erection may occur after the ages of ten, and during this period, the first ejaculation (thorarche) may involuntarily occur during sleep as a wet dreams (nocturnal emission). In both sexes, genital skin becomes sensitive to sexual stimulation, and teens may learn and practise masturbation or start to copy adult sexual behavior. Masturbation, at early puberty, could be a sign of sexual abuse; and without supportive parents, children may never talk about their abuse, and it can cause serious emotional and psychological problems both in boys and girls after puberty. Further, at puberty, the growth in height starts in girls before boys, and this may cause embarrassment to the boys. Equally, not all of the boys become tall or taller at the same period, which may cause further embarrassment and insecurity in short boys and girls and subject them to intimidation and bullying of their peers. Teens may also develop facial spots (acne vulgaris), which can be severe and causes further embarrassment and insecurity in both sexes. In addition, without family care, support, or guidance, the hormonal changes in this period can contribute to adolescents' unstable emotions, mood, temper, and behavior including violent behavior.

Their sexual practise at this stage depends on their family and cultural values and the conditioning of their sanity elements. Before puberty, boys and girls may start a shy "love" conversation; but after puberty, boys may develop voluntary or involuntary erection if they meet or speak with girls. Erection can occur during sleep with their sister or brother in the same bed; hence, it is important to separate their beds at puberty. Also, children tend to make friendships with similar-minded people; and after puberty, they may play or copy adult sexual behavior with their peers. This early sexual experience may affect (condition) their sexual orientation. Gradually, adolescents become more independent, curious to experiment or to prove their sexuality to their peers. In a conservative culture, adolescents' sexual behavior is usually secretive, platonic, or they may experiment sex with prostitutes or vulnerable people. In liberal culture, adolescents are more free to express their sexuality and sexual relationships. In general, this period is the most sensitive period in a human's physical, mental, and sexual development. Teens become curious about the biological changes in their body and in their peers, and childish behavior such as playing will gradually change into more acceptable behavior such as sports. Adolescents living in the cities are conditioned by the cultural and media norms and follow fashions to be physically and sexually attractive. Some of them feel fear or embarrassment of being unattractive, fat, skinny, short, or tall. Fat teens influenced by commercial media may start dieting, and some of them may develop anorexia or bulimia. Severe fear can also develop in adolescents who have doubt about their sexuality. In conservative cultures, homosexual boys and girls may suffer humiliation and bullying if they reveal their secret to their parents or friends. Equally, sexually abused children may suffer alone for many years or decades from shame, humiliation, which can affect their confidence and sexuality.

## Youth 18-29

There is no fixed year for leaving the adolescent period or behavior as people may maintain their "immature" behavior till their late forties. However, most adolescents finish their physical and sexual growth by age of eighteen; and in many countries, this age is regarded as the end of childhood and the beginning of personal independence and legal judgment. Many young people in this period start to face different types of fears and anxiety to build their own career, family, or goals in life. They leave childhood play, games, and dreams and face the adulthood reality. Without family support, failure to achieve their personal goals can create severe mental and psychological disorders, including extreme behavior such as homicide or suicide. Sexual behavior in this period would depend on the cultural values and the type of freedom available in the environment. However, in the last few decades, the Internet, especially sex Web sites and sex chat rooms, have made different sexual behavior available and accessible in most cultures. Although, during this period many people get married or have long relationships, after adolescence, many youths may try to explore different sexual activities with their lovers, peers, and people they meet at work, in the street, or through dating agents, pub, clubs, or brothels.

Fear of being sexually unattractive, fat, or thin may also continue in this period; and in large cities, some girls may become victims of commercial dieting, depression, obesity, or anorexia. Also, in conservative cultures, homosexual or cross-gender youth may face fear, humiliation, and abuse. Heterosexual youths may also fear the responsibility of marriage and commitment. The lifestyle of modern societies conditions youths to have high expectation and to be independent, selfish, and greedy. This has increased the rate of domestic problems, violence, separation, adultery, and divorce. The latter can lead to isolation, depression, and may involve children in custody problems. Separation or divorce can lead to vulnerability and sexual promiscuity and/or discourage divorcees from having another marriage commitment. Promiscuity and unsafe sex have been the main cause of the STI/HIV pandemic, which can further increase young people's anxiety, depression and fears. They may fear the risk of complications resulting from these infections, how others would perceive them, finding difficulty in discussing their sexual health issues with their sexual partners or with their general practitioners.

## Adulthood 30-65

There is also no fixed age when to enter adulthood as some people maintain their health and youth by fitness and healthy lifestyle while other may die before reaching sixty. Many people will build up their family life in this period and look after their children till their children reach adulthood. Their sexual lives depend on their belief system, sexual orientation, and cultural norms. People in monogamous relationships often engage in sexual activity more frequently than those who have several partners, but among married or cohabiting couples, the frequency of sexual intercourse tends to decline with age. There is also a decline in the number of STIs/HIV cases in this age group, and their incidence decreases with the age. The average age for women's menopause (the end of the menstrual cycle) is at fifty. Thereafter, the female hormone (estrogen) starts to decline and may lead to thinning and dryness of the vaginal walls or decrease in vaginal lubrication. These conditions can cause pain during intercourse, especially in women who stop their sexual activity after menopause for a long time, but it can be treated by local hormonal therapy and/or lubrication. Many women in conservative cultures start to lose their sexual interest while in liberal cultures older women are more free to express their sexuality and may continue their sexual activity till senility. Most men continue to be fertile, have morning erections and ejaculation till senility although their sexual desire and response decreases with the advancing age. Nonetheless, as both sexes grow older, they become more conscious about their age; and they may fear the rejection of being old, unattractive, and fear the inability to get work, or the inability to compete with young people. They may also suffer from embarrassment or rejection when they try to meet a new young partner.

## Elderly 66+

This period may start at fifties or seventies, depending on people's personal lives and the cultural perception toward elderly people. Although it may represent the golden age where most people are retired with rich life experiences and the freedom to enjoy life to its fullest, some of them become a victim of chronic illness, which disables their mobility and independence. Equally, the chronic illnesses may impair their sexual desire and activities. In addition, elderly people who lost their sexual partners may feel very depressed and less physically and sexually attractive compared with young people. Some of them may die before reaching this period; others who live alone may lose motivation to have a new partner. Further, some of them may fear diseases, dementia, and death or may feel unwanted or rejected in certain societies, especially if they have no family or friends. However, happily married elderly couples may live with their children or grandchildren and remain positive in their personal and sexual activities and in their contributions to society.

## Sexual Desire in Men

Sexual desire is an instinct and is influenced by the competency of the sanity elements, five senses, and ANS reflex. Penile erection in men is a voluntarily and involuntary process during awake time and is involuntarily during sleep. After puberty, penile erection, wet dreams (nocturnal emission) which include orgasm, and ejaculation can occur during sleep without any physical or muscular efforts. The conditioning of the sanity elements, especially belief system, feeling and emotions at early childhood has a major influence on human sexuality. Sexual response in men is also influenced by their medical and psychological status. Certain medications and illnesses such as anxiety, depression, vascular diseases, neurological diseases and endocrine diseases, and tumors can cause failure of erection or impotence. Uncontrolled diabetes mellitus is a common cause of impotence.

During adolescence, boys start to be conscious about the growth of their body and genitals. Some of them become concerned about the size of their penis. However, ejaculation in men is not influenced by the length of the penis, but by the ejaculation spot or "e-spot," which is located just below the head of the penis. Hence, ejaculation can be achieved even with a small penis. Metaphorically, this area represents the "existence" spot as without its stimulation; human generation may have not been existed. Without erection and ejaculation, men can't consummate a marriage. The latter may cause anxiety and shame to the groom, and in certain cultures it may lead to divorce and loss of the groom's money and dignity. In contrast, after many years of marriage, impotence may cause anxiety to the wife as she may feel unattractive.

Potentially, any healthy, mature man can make voluntary penile erection at any time by fantasizing, and without the need for a sexual object. However, not all men are able to develop sexual fantasy or voluntary erection, especially anxious men. Anxiety and fear from failing to achieve erection and ejaculation can be very distressing to men after marriage, and in some men, it may last for months. However, by sexual experimentation with their wives, their

ANS erection reflex condition toward their wives and erection may become spontaneous whenever they think about or meet their wives. Even in homosexual men, who are unable to develop erection toward women, repeated sexual experimentation with women can condition their ANS erection reflex toward their female partners.

In general, sexual arousal in men starts after puberty and passes through five main stages or periods as follows: *first stage* or *the sexual desire period (libido)*, during which men feel desire or appetite for sex but without erection. The erotic feeling can last a few seconds or for many hours and is influenced by many factors such as age, competency of sanity elements and senses, self-confidence, belief, mood, interest, physical attraction, place, time, previous sexual experience, and mental and medical health. Men may feel erotic when they visit certain places or meet certain people or are alone with their wives, but their feeling may or may not progress into the *second stage* or *the sexual arousal period*, during which the penile erection can occur voluntary or involuntary and even without the sexual desire stage. The erection develops as a result of the stimulation of the parasympathetic part of the ANS after receiving sensory stimuli (e.g., watching, hearing, thinking, or fantasizing of erotic objects or sexual stimulation of the genitals by self or by a sexually attractive person. The parasympathetic part of the ANS causes dilatation of the blood vessels in the penis, which helps to pump blood into it within seconds. Once filled with blood, the penis becomes firm and erect from a flaccid position. The erection may develop within seconds and may last for few seconds or minutes. The involuntary erection can cause severe embarrassment for men in public place as they feel unable to stop it while talking or meeting a sexually attractive female.

In addition to erection, during this stage, the skin of the scrotum thickens, and the scrotum is pulled up closer to the body; and both men and women develop increased muscle tension and increased heart and respiratory rate and blood pressure. The persistence of penile erection for more than thirty seconds represents the beginning of the *third stage* or *the plateau period*, during which sustained penile engorgement, muscle tension, elevated heart, blood pressure, and respiratory rates continue; and the head of the penis (glans) swells and a clear sticky secretion comes from the end of the penis called preejaculatory fluid or precum. This fluid nourishes the sperm and may contain active sperm capable of impregnating a woman. Also, in this stage, both women and men may develop vasocongestion or "sex flush," which usually starts on the upper abdomen and spreads to the chest. The plateau stage may last for a few minutes and cease, especially in a newly married anxious man who is under tension of reaching orgasm and ejaculation. Improving feeling by improving the perception of the five senses (e.g., nice holiday, flowers, nice bedroom environment, nice smell, and hearing nice music, eating delicious meal, and manual stimulation of e-spot) may help to ease tension in anxious men. Some people may try to fake orgasm to please their partners. This is more easy in women as men can't fake orgasm without erection. If the plateau stage progresses, men may enter into the *fourth stage* or *the orgasm and ejaculation period*, which induced by the sympathetic reflex of the ANS and lasts less than one minute and during which around five to ten involuntary rhythmic contractions of pelvic musculature occurs. Breathing rate, pulse rate, and blood pressure increase dramatically during orgasm; and general muscle contraction may also occur. In both sexes, orgasm or climax causes pleasurable sensation. In men, orgasm is associated with ejaculation of semen, which results from contraction of the vas deferens, seminal vesicles, and prostate, sending seminal fluid to the urethra (see Figure 1). However, men may be able to prevent ejaculation by pressing on the proximal part of the penis, and this leads to retrograde ejaculation or the return of the semen backward in the pathway. Retrograde ejaculation can also occur after prostatectomy. In contrast, a few men develop premature ejaculation or ejaculate within thirty seconds of penetration and before the orgasm of their partner. This condition is more common in young men and may cause discomfort to their sexual partners. The treatment of premature ejaculation requires relaxation while practising sexual intercourse and gentle stimulation of the e-spot. The latter can be achieved by gentle repetitive penetration of the distal part of the penis (e-spot) in the vaginal orifice and withdraw the penis before reaching orgasm. Gradually, the affected man will feel more control in the stimulation of his e-spot and his orgasm. The ejaculation in men may spurt two feet, and it contains jellylike whitish fluid called semen. Its volume may vary from 1.5-5 ml and decreases with frequent performance of sexual intercourse and with advanced age. After ejaculation, erection ends, and the penis becomes flaccid within a minute; and men feel relaxed, tired, or sleepy and pass into the *final stage, the refractory period or recovery period,* which represents the period of time required for a man to have another erection after ejaculation. Hence, during the period, men are incapable of having another erection. The erection subsides, and the penis returns to its normal size within a few minutes so are the scrotum and testes. The duration of the refractory period may vary from five to thirty minutes, but it may last many hours in elderly men.

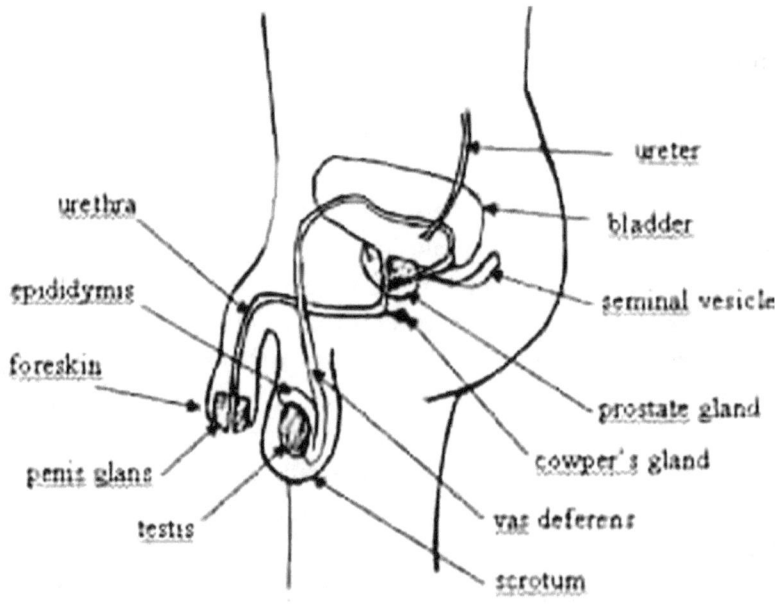

Figure 1: Drawing of male sex organ

## Sexual Desire in Women

Apart from ejaculation, sexual response in women includes the same sexual stages as men. However, woman represents the land for seeding human generations while man is the seeding farmer. Without women, there would be no human reproduction or civilization, and men would be unable to become fathers. Women are also more emotional and tearful than men. They have estrogen hormone and no "testosterone urge" or involuntary erection. The sexual arousal stage in women may take a longer time, especially in the virgin. Also, orgasm in women could be multiple or may never be reached. The latter is common if the sexual male partner is quick and does not have time or empathy for women's sexual needs. Nonetheless, even with a loving male sexual partner, not all women can achieve orgasm during each sexual intercourse. Just like in men, women sexuality can be influenced by cultural and media conditioning, at early childhood. The media focuses on the visual attraction or image, and it can condition a woman's personal behavior including her sexuality. Despite women's beauty, the media conditions many women to be victims of makeup or fashion. They may feel unattractive or insecure without makeup. Equally, the media betrays fat woman as sexually unattractive, and this may affect her self-confidence to find a sexual partner. In addition, cultural values can also affect women's sexuality. Women working in male-dominating fields may lose their femininity. The latter is the most powerful magnet that can attract any men regardless the beauty or size of the women. Fat feminine female can sexually stimulate any men better than any beauty queen who has no femininity. In addition, orgasm may never occur in women who live in very conservative cultures and have no information about the sensitive part of her body and/or have fear from aggressive men, STIs/HIV, pregnancy, or abortion. The virgin bride, with no sexual experience, may develop severe anxiety and fear of being penetrated and losing her virginity for the fist time by a hostile man without prior stimulation or foreplay. Women are physically more delicate than men and usually need gentle stimulation of the sensitive parts of their body such as the lips, the nipples, and the clitoris. The latter is a tiny penislike organ situated in the upper corner of the vagina (see Figure 2). It represents the e-spot in the male. However, even with stimulation of the clitoris, reaching orgasm in women can be very difficult and need frequent sexual experimentation with her sexual partner. Further, women may achieve clitoris stimulation more easily by taking the top position during sexual intercourse. Such position can give women more self-control on stimulating the clitoris during sexual intercourse. Nonetheless, couples may need to experiment with different positions and enjoy the fantasy of sex. The latter include focusing the thinking on erotic pleasure while anticipating the orgasm during penetration or during stimulation of the clitoris. The head of the clitoris (glans) is just like the glans penis in men and can be very sensitive to direct touch. Other sensitive areas include the inner surfaces of the labia minora and the first part (distal part) of the vaginal wall. Considering the e-spot in men, slow and frequent penetration of the distal part of the erected penis in the distal part of the vagina (vaginal orifice) may achieve better chance for reaching orgasm in both sexes than forceful sex. Breast and nipple sensitivity tends to be high in women, but not all women or men find breast caressing arousing. It might also be helpful if a woman with orgasmic failure tries to explore the sensitive part of her body alone, and then she can tell her partner how she would like to be touched. Some experts believe that the G-spot, which is supposed to be in the vagina or in the rectum, triggers orgasm. Other experts think that the G-spot is a myth and creates confusion, anxiety, or obsession in both sexes. More importantly, couples should try to keep sex pleasurable and to enjoy it. This may be difficult if one partner or both partners have personal or domestic problems. Also, women brought up in a strict religious culture may develop fear or guilt from exploring their sexuality. Other factors that might impair sexual desire in women include stress, fatigue, depression, pain, and fear and using certain drugs or hormones, unpleasant previous sexual experiences, and loss of interest in a partner. In infertile couples who are trying to conceive, the anxiety of following instructions of having intercourse around ovulation time can be very stressful and impair sexual arousal and orgasm in both men and women.

The *libido period* in women is more common during the middle of their cycle when the estrogen is released in greater quantities to prepare women for pregnancy. If sexually unaroused, the labia majora in women lies close to each other while the labia minora are usually folded over the vaginal opening, and the walls of the vagina lie against each other. During the *sexual arousal period*, some changes happen in women as result of the blood pumping into the sex organs, causing enlargement of the breast and the clitoris; and involuntary clear, sticky vaginal discharges start to appear, causing fluids to seep through the vaginal walls to produce vaginal lubrication. Just like the precum in men, the clear vaginal discharge is an important sign of sexual arousal in women. During the *plateau period*, the female sex organs enlarge further, the clitoris becomes erected, and its tip (glans) becomes larger and harder than usual. Muscular contraction around the nipples causes them to become erect, and the labia majora spread apart while the labia minora swell and open. The upper two-thirds of the vagina expand, and the cervix (neck of the womb) and the uterus pull up, helping to accommodate the penis during sexual intercourse. After sustained sexual excitement, the breasts continue to swell, the lower third of the vagina swells, the uterus enlarges, the clitoris retracts into the body, and the labia majora darken. The *orgasm* in women usually starts in the clitoris and spread to vaginal wall and the uterus, causing around three to ten strong contractions of the vaginal wall and the uterus within less than a minute. These contractions cause intense sensation more than the tingling or pleasure that accompanies sexual arousal. Unlike men, women can develop multiple orgasms during a single sexual intercourse and without passing through the refractory or resolution period. However, absence of orgasm is more common in women. If orgasm occurs, then women enter the *refractory period*, which may take five to thirty minutes; and during which, there is a reduction in muscular tension, the sex flush disappears, and the sexual organs return to their original size before the aroused state. The muscles of the vagina become relaxed, and the neck of the womb dilates, which helps the sperm to enter into the womb. Apart from orgasmic failure, few women may develop vaginismus, which is a spastic contraction of the lower third of the vaginal wall. This can result from fear and anxiety, and it can prevent penile penetration. Also, painful intercourse (dyspareunia) can occur in both women and men. The common causes of painful intercourse in women are dry vagina due to lack of sexual stimulation, menopause, and vaginal infection, especially vaginal thrush (candidiasis). Also, libido may be impaired in women during pregnancy and after the birth of their child (puerperium) as a result of hormonal changes, which can cause dramatic physiological, psychological, and sexual changes and can last many months after delivery. After birth, women may suffer from fatigue, vaginal discharge, vaginal bleeding, insufficient lubrication, fears, and stress of awakening the infant or fear of being nonattractive to her partner. Some women may under go vulval surgical incision (episiotomy) to assist baby delivery. Such incision can cause painful intercourse or discomfort for many months.

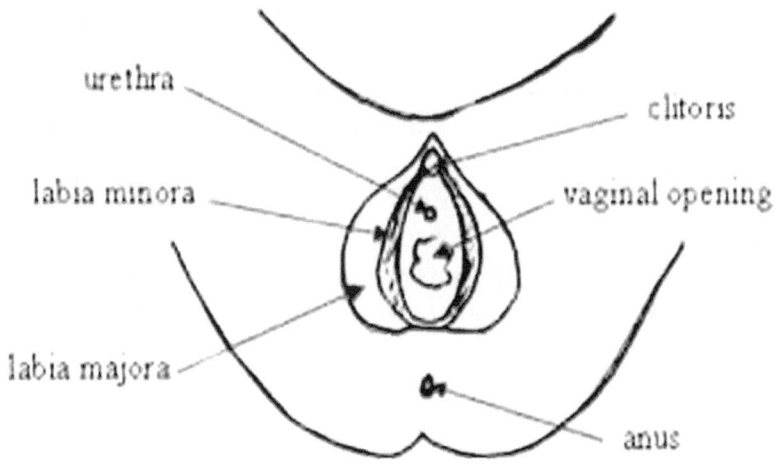

Figure 2: Drawing of female sex organ

## Masturbation

Masturbation is a learned (conditioned) sexual behavior during which a man or a woman voluntarily stimulates his/her sex organ with the intention of having sexual pleasure and/or orgasm (and ejaculation in men) but without penetration. It requires thinking or fantasizing about erotic object while manually rubbing the genitals (penis in men or the clitoris in women); hence, it may be impossible to achieve masturbation without hands. Some animals may masturbate or enjoy rubbing their genitals; and before puberty, children may have curiosity to play with their genitals, but without arousal or erection.

However, masturbation in children could be a sign of sexual abuse. In general, masturbation is not an acceptable behavior in most cultures; hence, it is done secretly, and it is difficult to estimate its prevalence in each culture. Studies in liberal societies suggested a high figure of masturbation among men and women of all age groups, including married couples. Nonetheless, practising this habit is influenced by many factors such as time, age, culture, media access, loneliness, boredom, and sexual conditioning. In a very conservative culture, many men and women remain virgin while in liberal culture, adolescents may start masturbation after puberty. As masturbation is associated with temporary pleasure, many people become addicted to this habit, especially when they are alone or feel bored. There are misconceptions, gossips made around masturbations, but without evidence. Nonetheless, although masturbation is safe from illness and STIs, it would not replace human interaction; and men and women can live normally without masturbation. Repeated masturbation can reduce the ejaculation volume and the ability to perform further sexual act and may also make men feel tired and/or affect their sleep rhythm. Furthermore, some men become addicted to practise this habit on daily basis and may feel guilty and tired from practicing the same rituals. Masturbation can be dangerous if it is done in public places, while driving a vehicle or operating heavy machinery. In contrast, masturbation can save a man from rape charges if he has poor willpower to control his sexual desire. It can also save him from acquiring potentially fatal STIs if he missed using a condom with a casual partner.

# Human Sexuality

Sexuality represents the sexual side of the human being. It includes all the secretive mental and sexual activities. Human desires including sexual desires are "self" wishes and are linked to their sanity elements, especially beliefs, emotions, and feelings. The latter depend on the competency of the five senses, especially vision and hearing. These two senses are important for human attraction, discrimination, curiosity, searching, learning, or conditioning. Human feeling is influenced by the sensory stimuli present in his environment. Metaphorically, if a man was born alone in the moon, he would not think or develop desires for a person or an object he couldn't see, hear, or think. Similarly, if there is only one type of apple in the moon, he would not feel greedy to look or taste other types of apples. However, the presence of many types of apples can make him curious and greedy to taste all types of apples, and he might end eating the most delicious one. Similarly, the presence of many women in the moon can make him curious and greedy, and he might choose the most desirable one. The more temptations people "see" or "hear," the more they develop desires and/or curiosity to try them. Nonetheless, on earth, human behavior including sexual behavior is restricted by cultural traditions. Although appetite for food is similar to appetite for sex, the former is acceptable in public places, while the latter is not; hence, most of human sexuality are secretive. In addition, unlike animals, humans' sexuality is linked to their sanity elements, which are influenced by cultural conditioning. The conditioning of people's belief system and emotions at early childhood can influence their sexual desires. Both food and sex desires cause pleasure, and any behavior associated with pleasure can lead to obsession and can condition human ANS reflex. The latter can induce salivation during watching or thinking about delicious foods and induce sexual arousal and discharge (precum in male and vaginal discharge in female) on watching or thinking of desirable sexual objects. The link between human sanity elements during thinking or watching delicious food or desirable sexual object is as follows:

1.  **Thinking**: Thinking of foods or sex (desires) may occur spontaneously, by fantasizing (memory) or after watching or talking about them. Repeated thinking about desires can lead to obsession and condition the ANS reflex to certain objects or act (behavior).
2.  **Memory**: (Thinking) can trigger memory of pleasure (desire) and its (emotions), which can push people into an action (behavior).
3.  **Feeling**: Seeing delicious food (desire) can make people feel the food in their mouth (think). Watching sex movie can make people feel the sexual penetration and can trigger their ANS sexual arousal (behavior).
4.  **Understanding**: Understanding how to get (thinking), what to do (behavior), and where to go to have safe (belief) pleasure of food or sex (desire) encourage people to practise (behavior) their desires.
5.  **Belief**: People's belief system (trust) can determine (thinking) whether to eat delicious food (desires) or not. It also determines whether to stay virgin or to have sex with certain person(s) (behavior) and in certain circumstances (learning).
6.  **Emotion**: Positive emotions (e.g., love and happiness) can trigger (thinking) people's appetite and may increase their (desires) for food and/or sex (behavior), while negative emotions (e.g., depression) can impair their desires for food and/or sex.
7.  **Learning**: Learning can condition all sanity elements including people appetite (desires), fantasy (thinking), and instinctive reactions (feeling and emotions) and condition their trust (belief) with themselves (willpower) and with others (behavior).
8.  **Desire**: People (think) of their desires and remember (memory) the previous pleasure experience and react (feeling, emotions, and behavior) according to their (belief system).
9.  **Will**: Willpower correlates inversely with (desire). People's willpower reduces if they have high desire and obsession (thinking) to certain food or sexual objects.
10. **Behavior**: If people have high (desire) to a certain object and high trust (belief system) in getting it safely and/or poor (willpower), they would become curious or obsessed (think) and try to have it (behavior).

However, unlike food, practising sex is not socially acceptable and needs sexual partner(s). The latter is not always accessible or acceptable without marriage; hence, most human sexuality remains secretive. Further, sex desire correlated positively with belief and love emotion. People tend to have sex with whom they trust and love. Nonetheless, most pleasurable behaviors in humans are liable to obsessions, compulsions, or addiction especially in people who have poor willpower. The latter may lead to obsession and addiction to practice sex without love emotions. Equally, human learning can condition their belief system and desires. After weaning from milk, infants start to taste different kind of foods and develop desire to certain foods. For example, the pleasure of eating sweets can condition their ANS reflex to cause salivation on seeing, hearing, or thinking of sweets. Gradually, many children become obsessed about sweets and develop compulsion to buy, steal, or eat sweets even if they are not hungry. However, if the infants have never been exposed to sweets, they would not develop desire to them. Similarly, after puberty, boys develop penile erection while watching, hearing certain sexual behavior or movies. The pleasure and the sexual excitement of watching sex movies can condition their ANS reflex to cause sexual arousal to certain person, objects, or behavior. They may fantasize, become curious or obsessed about these sexual objects or behavior. The curiosity and pleasure of sex can push them to practice sexual behavior or masturbate whenever they feel bored, stressed, or lonely. In contrast, a blind and deaf person would not be able to do the same. They may feel scared to explore their sexuality or masturbate as they can't see or hear who is present with them in the room. Nonetheless, the conditioning of the people living in the cities is different from those living in a remote jungle. Children born in the cities develop different interests, attractions, and needs. They see and hear different types of sensory stimuli and develop different perceptions, expectations, and belief systems. Emotional depravation such as strict discipline or lack of discipline and support can have major impacts in conditioning of children's belief system, willpower, desires, feeling, or perceptions. According to their environment and to their instinctive needs, after infancy, children start to perceive themselves as weak, strong, lucky, unlucky, rich, poor, rejected, wanted, confident, shy, masculine, or feminine. The human family is the seed of the personality. A supportive relationship between a father and the male infant can give the child the sense of belonging and a strong male gender identity. It can condition the sanity elements of the male child to copy his father's emotions, beliefs, expressions, interests, ways of communication, and masculine behavior. The same can occur between a female child and her supportive mother. Without supportive parents, children can be conditioned to any sensory stimulation that exists in their environment or in the media. The latter conditions people's sanity elements and desires for certain images. Children learn from an early age the characteristics of beautiful or desirable people. They learn that very fat or very short people are regarded unattractive or funny. This can condition children's perception and belief system or expectation. They may even make jokes about fat or short people. After puberty, the conditioned sighted boys may become emotionally and sexually attracted toward beautiful thin girls while conditioned sighted girls become emotionally and sexually attracted toward tall masculine men. Fat girls may become conscious about their body sizes and may seek dieting to achieve the "attractive figure." Similarly, boys may seek bodybuilding or certain sport activities to attract their peers' attention. In addition, the hormonal changes after puberty have an impact on children's sexual and emotional development. Without emotional support, adolescents can be the most critical period in human life. The hormones do not only increase the height of children but also change their emotions, mood, or perception. During adolescence, boys and girls start to compare their body growth with their peers and develop different perceptions and expectations. They may perceive themselves as weak, strong, fat, thin, tall, short, masculine, feminine, attractive, or unattractive to their peers or to any person attracting their attention. Adolescents, who perceive themselves as unattractive or perceived by their families or their peers as unattractive, may feel low, unconfident, insecure, or shy to meet people compared with adolescents who perceive themselves attractive. They may also feel emotionally and sexually vulnerable to people who they perceive as attractive. The negative emotions in adolescents may evolve into obsession, guilt, and/or depression. These conditioned emotions are more common in teens living in the cities compared with the teens living in jungles. Without support, the negative emotions can influence teens' belief system and sexual orientation and make them vulnerable for love and/or attention from people who they perceive as physically and/or sexually attractive. Gradually, their conditioned belief system conditions their ANS sexual arousal reflex toward people who they perceive attractive and determines the way they perceive their sexuality and the way they perceive other people sexuality.

Furthermore, people's sexual behavior is influenced by their cultural values, faiths, and laws. Without law, faith, or cultural values, human sexuality can include all types of taboos or paraphilia. However, most people are conditioned to be heterosexual, which is the "acceptable" sexual relationship in most cultures and faith. Other sexual orientations have been "unacceptable" or forbidden, hence become secretive. Nonetheless, the sexual revolution after World War II, the inventions of contraceptive pills, and the lifting of the ban on abortion in the liberal cultures have changed the belief system of many men and women living in these cultures; and they start to explore and express their sexuality without fear or better than before. Moreover, in the last decade, the global media, including the Internet, has become the new "heroin" for people's daily sexual stimulation and excitement. The sensory culture (visual and auditory) of the media can trigger sighted people's appetite for sex. Many sighted children living in the cities are able to see or watch sex, and they may copy adult sexual behavior even before puberty. The media can also condition people's belief systems and can make *unacceptable* sexual behaviors *acceptable* or accessible. The sex media can undermine the moral and ethical values and trigger teens' appetite for all types of sexual behavior including sexual abuse and rape. Without family discipline and/or support, teens living in large cities can copy different types of sexual behaviors and acquire wide physical and sexual attractions. The conditioning of the sanity elements can make humans more obsessed or vulnerable to their desires than animals. Animals do not have sanity or develop addiction to sex, food, alcohol, smoking, illicit drugs, Internet, gambling, or have desire to kill for money or racism. They also have no preference for height, weight, waist size, bank account, eye color, or hairstyle of their sexual partner; their sexual behavior is an inherited *unconditioned* impulse. Usually a male animal develops sexual impulse or *urge* toward a female animal from the same species. However, sexual urge can occur between animals from the same sex or between animals from different species, but without emotions or relationship. The human beings are the only conditioned creatures able to be explore and experiment with their desires and able to commit extreme sexual behavior with or without emotions. They are able to have sex for money, and for many days, with multiple partners and in different positions with or without emotions. Prostitution is a product of a selfish civilization, and it is rare among people who grow up and live their life in a jungle. Hence, changing people's sexuality is difficult as they grow older, and it may require changing their belief system and/or environment. The table below shows the difference between animal and sane human's sexual response.

|   | Animal Sexual Response | Human Sexual Response |
|---|---|---|
| 1 | Inherited (unconditioned) | Learned (conditioned) |
| 2 | No sanity elements | Related to sanity elements (emotions) |
| 3 | Not related to physical looks | Related to physical attraction, looks |
| 4 | Not liable to obsession | Liable to fantasy, obsession, addiction |
| 5 | Instinctive impulse (irresistible) | May stay virgin or promiscuous for life |

| 6 | Starts after puberty | Conditioning usually start before puberty |
|---|---|---|
| 7 | Limited behavior | Can be extreme |
| 8 | Lasts for a few minutes | Can last for hours or days |
| 9 | Ends by orgasm or ejaculation | Ends by leaving the sexual fantasy/object |
| 10 | Not influenced by cultural factors | Influenced by drugs, medical or mental problems |
| **Difference between Sexual Response in Animals and in Sane Human** | | |

## Sexual Orientation

Sexual orientation refers to a person's sexual attraction to certain gender or sexual objects. A person who is sexually attracted to the opposite sex is recognized as heterosexual or straight, and a person who is sexually attracted to the same sex is recognized as a homosexual or gay while a person who is sexually attracted to both sexes is called bisexual. Asexual is a person with no sexual desire.

## Male Homosexuality

The term "homosexuality" refers to same-sex erotic attraction, and currently males and females who practise homosexuality are commonly known as gays and lesbians, respectively. Also, men and women who are attracted to both sexes are known as bisexuals. Historically, there were extensive writings about homosexuality. In ancient Athens and ancient Greece, sexual attraction between men was the norm. However, since the biblical story of Sodom, many religions and cultures forbid homosexuality; hence, it remains as a secretive behavior. During Queen Victoria's time, buggery laws were aimed specifically at male same-sex sexual activity and did not target or address female homosexuality. Until 1973, homosexuality was categorized as a mental illness in many psychiatric textbooks. Thereafter, it was declassified from the *Diagnostic* (DSM-III), which is an American handbook for mental health professionals. In 1997, the Church of England approved a motion stating that practising homosexuals could become lay members of the church, but not priests; and by 2003, the Supreme Court in the USA reversed its 1986 decision and voided antisodomy laws in many states. Nonetheless, the homosexual relationship is still regarded as an *unacceptable* behavior in many conservative cultures. There are also social limitations for *acceptable* man-to-man behavior. For example, in the West, men are not allowed to greet other men by kissing in public like women do, and there is also color phobia. Men are not allowed to wear pink color. In many conservative cultures, men are expected to behave in a masculine way, as they may be perceived homosexual if they behave in a feminine way. Further, although promiscuity is common in both homosexual and heterosexual men, the latter is more acceptable than the former. Many men are proud to express their multiple heterosexual experiences in front of their peers, but scared to express a single homosexual experience or "feeling." Moreover, in certain cultures, people regard the passive homosexual man (a man who receives anal sex) as a cheap or sick person, addicted to anal sex, while the active homosexual man (a man who performs anal sex) is regarded as a strong masculine man. They may refuse to talk or work with a passive homosexual man, and/or they may feel shame to introduce him to their friends and families. Some parents may fear that their children may become homosexual if they were mixed with homosexual men; and many cultures regard homosexuality as more a *shameful* behavior than fornication, adultery, racism, discrimination, deception, theft, corruption, abuse of human rights, or homicide. Nonetheless, human behavior, including his sexuality, is a product of his environment (family, culture, and media); and experts are still struggling to prove whether homosexuality is due to genetic or hormonal imbalance in the embryo (Ref: 2). However, as I mentioned, environmental factors play a major part in all humans' behavior, including their sexuality. Homosexuality is rare in a man living alone in the moon or among people living in a remote jungle. Its prevalence increases in the larger cities, and there is a difference in urban people's sexual behavior. For example, homosexuality and buggery. The former includes feeling of love and attraction of a man to another man while the latter includes the physical submission of a man to another man for sex with or without love. There is also a difference between homosexuality and promiscuity. Both heterosexual and homosexual men can be promiscuous, but homosexual men may never have physical sex with men. Homosexuality can be only a feeling or emotions. Unlike animals, potentially all men can be conditioned to practise buggery with men and women with or without love. However, sexuality in humans is influenced by their sanity elements, especially belief, feelings, and emotions. Love emotion can make infants attached to their parents and can make parents or two strangers emotionally attached to each other for life. Human sexuality may start as emotional needs in early childhood and evolves into emotional and sexual need after puberty. Children are born with different emotional needs, interests, and physical activity; but their *luck* is totally dependent on their environment.

Some of them may have a higher emotional threshold than others. Unlike children living in the jungles, children living in the cities undergo more complex conditioning. They are exposed to more temptations and develop different needs and expectations. The father represents the masculine role model in the family and is the main financial support for a male child. A supportive and loving relationship between a father and his son can condition the son's behavior toward his behavior and give the son a male identity. Without a loving, caring, or supportive father, the sanity elements of a child enable him to fantasize to fulfil his emotional depravation. The boy may start to look for love, trust, and support from others to fulfil his needs. Homosexuality may start as love and trust between two boys or two girls who feel comfortable to share love, sex, and secrets together. A careless, dominant, selfish, ignorant, abusive, or boring father usually fails to build a trustful relationship with his son. Equally, a careless, dominant, selfish, ignorant, impulsive, or hysterical mother may fail to build a trustful relationship with her son; or she may condition the child to hate his father or encourage him to perform feminine activities. This may undermine her son's masculine behavior, and with the lack of father support in the family, the child may become conditioned to his mother's feminine behavior and emotions. After puberty, boys develop spontaneous erection, and it is not at all uncommon that early sexual exploration including sexual penetration may happen with other members of the same sex. The pleasure of sex can push some children to repeat the same sexual experience with the same sex before they are able to get married. The repeated early sexual experience between two boys or two girls after puberty can condition their perception and their ANS sexual arousal reflex (erection in boys) toward the same sex. Even wet dreams (nocturnal emission) in homosexual males may remain exclusively homosexual dreams. Conditioned boys and girls may start to perceive and expect (belief) sexual pleasure and excitement from the same sex more than from the opposite sex. However, although conditioned ANS reflex (erection) in boys with homosexual experience may remain toward men only, not all of them become homosexuals or sexually attracted only to same sex. The continuation of homosexual behavior depends on the conditioning of adolescents' sanity elements, especially their belief system and not on the conditioning of the ANS sexual reflex. Sexual arousal (erection and precum) in men is irresistible involuntary neurological reflex of the ANS and can occur within seconds of meeting, thinking, or watching a desirable person; but it is not always correlates with their love emotion and belief system. Similarly, seeing delicious food can cause salivation, but the belief system and willpower enable a person to reject eating food even if they are very hungry. Boys with homosexual experience may develop erection only toward men, but not all of them become homosexuals as their belief system is conditioned differently in each environment. Adolescents with emotional and sexual needs who find more attention, attraction, love, support, trust, sexual pleasure from men and not from women may continue their homosexual behavior, especially if they live in a liberal culture, while adolescents with religious beliefs who live in conservative cultures may fear sin or social stigma. They may also feel or perceive themselves sexually *abnormal* and suffer severe distress and guilt feelings. The latter can be the most traumatic and fearful experience in an adolescent who is supposed to have a normal erection toward women and not toward men. In the conservative culture, such an adolescent may suffer alone and feel fearful to talk about their feelings to their parents, family, and friends for years or for good. In addition, the emotional (love) needs may increase after puberty with the increase of the sex hormones, and this can cause more emotional and sexual repression and frustration in adolescents. Some adolescents with homosexual experience may feel unable to *give up* or control their sexual attraction and/or love emotions toward certain men and may commit suicide to escape the mental suffering. Nonetheless, not all of adolescents with homosexual experience have similar belief systems or conditioned in the same environment. Some of them develop a fixed belief system that they were born homosexual; others develop a flexible belief system, which enables them to experiment sex with men and women or be bisexual even after marriage. In general, one or more of the following acquired factors may influence male homosexual behavior:

1. Hostile father, mother, or older brothers
2. Lack of father's love, support, or guidance
3. Negative women images in the family
4. Dysfunctional family and emotional depravation
5. Parents treating their boy as a girl
6. Religious fear from having sexual relationship with women
7. Lack of feminine or attractive girls in the same culture
8. Lack of cultural mixing with girls
9. Repeated rejections by girls
10. Low self-esteem or confidence to attract girls
11. Child sexual abuse or humiliation to his manhood and confidence
12. Copying peers' behavior or peer pressure
13. Access to sex with men more than women

14. Lack of fear or inhibition (e.g., excess alcohol or use of illicit drugs)
15. Financial needs or greediness for money may push few male for prostitution
16. Curiosity to copy the sexual pleasure shown in the media (e.g., oral and/or anal sex)
17. Desire to be close or to build allies with powerful or rich men through sex
18. Cultural norm, conditioning or tolerance to homosexuality
19. Unable to marry (e.g., for religious reason)
20. Superiority, selfishness, and power abuse

The last example is more common in prisons, military services, and schools where a physically powerful man or teen sexually abuses a weaker man or teen to be proud of their manhood in front of his peers. Such abusers usually regard themselves as heterosexual and are proud to practise incentive anal sex and abuse or humiliate his victims. The abusers are usually conditioned by their selfish parents to be selfish, dominative, or proud about their physical power or hostility. As they grow older, abusive children with physical power feel insecure unless they dominate or abuse other people who live or work with them. The irony is that some insecure men and women feel sexually attracted to such physically powerful, abusive men, and they may build up allies with them through sexual relationship. Nonetheless, homosexual men are just like heterosexual men in that they have different sexual behavior and loyalty to their partners; and there are different types of homosexual practices, ranging from nonpenetrative sex or mutual masturbation to aggressive penetrative, unsafe sex with multiple partners. As sex causes pleasure, it may lead to obsession and compulsion; and just like some heterosexual men who are obsessed with certain types of women or certain parts of women's body, some homosexual men may also develop obsession toward certain types of men. The sexual obsession could push some men to practice masturbation in public toilets or have sex with stranger(s). The latter behavior can be dangerous as many homosexuals may acquire STIs or HIV if they do not use a condom and some of them have been abused, raped, mugged, and killed by strangers or serial killers.

## Female Homosexuality

Females are more smooth, gentle, and peaceful than men, at least historically and biologically. Most of crimes, abuses, rapes, massacres, and wars were committed by males. Females show more emotions, especially toward their children, and react differently to emotions. They are relatively more expressive and sociable than men. They do not have the testosterone urge, voluntary or involuntary penile erection, during the day or the night, which makes them less of a "sex animal" than men. They can live and sleep together in the same bed for many years without sex. However, women have been conditioned in the male-dominant culture. In the urban area, women are more attracted to romance, flowers, fashion, and makeup compared with women living in the jungle. With the dominant male world, it is not surprising if some women dislike male superiority, pride, urgency, intimidation, harassment, and abusive behavior. However, not all women who dislike men become lesbians, and not all lesbians have similar sexuality. Many factors that lead to homosexuality in men (above) can also lead to homosexuality in women. In addition, unlike men, sex in women is not a ten-minute job of penetration and ejaculation. Most women become sexually aroused by sensory perception or feeling. Love, trust, care, romance, flowers, gifts, and prolonged foreplay with sexual stimulation especially to the clitoris can help women to *feel* erotic. Unfortunately, most men are quick and ejaculate within ten minutes. This may not give enough time for a woman to develop orgasm. Hence, some women are curious to explore sexual pleasure with other women who understand her sexual needs and the sensitive parts of her body. Even after marriage, some women may become curious to have a homosexual experience to explore the pleasure of sex. Equally, some women become curious in copying lesbian sexual pleasures they watch in the media and feel more comfortable and safe to experiment sex with a nice, soft woman than with an unpredictable, hostile man. Also, some women may try to develop a close emotional relationship, bonding with their female friends by sex, especially if they have had a bad experience with men or their male partners do not spend enough time to fulfil their emotional and sexual needs. Other women may feel more attracted to the physical beauty of the female body, lack of facial hair, kindness, softness, and feminine behavior in women more than the aggressive behavior in men. Women brought up by an abusive father or brother(s) may prefer homosexual experience, especially after having unsuccessful multiple heterosexual relationships with abusive, selfish, or unfaithful men. The superiority, arrogance, and continuous harassment of men to women at work, on the streets, and in public places may condition some women to hate male pride and dominance. In addition, some women may find that the male penis is very ugly, smelly, or repulsive; and they may prefer to use colorful sex toys such as dildos or vibrators, which do not have the shape of a male penis for sexual stimulation. Such toys are safer to use and can fulfil their sexual desire without the hassle of the penis, which could cause pregnancy and STIs. Furthermore, some women exposed to sexual assault and rape at childhood feel fearful to be near a man or to be touched by a man, and they may need a long time of support to overcome the fear and the distress of sexual abuse. Their fear may increase at adolescence, and they may feel more comfortable to share sex with a caring, kind female rather than with a hostile, selfish man.

## Gender-Identity Disorder

Gender identity disorder or transgender is the feeling and/or the desire of being the opposite gender. This may include copying feminine or masculine behavior, cross-dressing, transvestism (TV), and transsexuality. Transgender is more common in men and usually starts at early childhood. A boy may feel more comfortable to behave like a girl and have girl's toys. A transgender person is not always homosexual, some of them are married, and some heterosexual men prefer a transgender person. Genetic and hormonal factors during pregnancy were blamed as a cause for transgender, but have not been proven. There has been no large cohort study conducted on the human embryo or in infants to measure the relation between genes, sex hormones, and the feminine or masculine behavior of the children. With exception of a very rare congenital condition called intersex in which the children are born with ambiguous genitalia (see "Embryo -0"), acquired factors play a major part in gender identity disorder. Although some children are born with variable feminine and masculine activities and behaviors, not all of them develop gender identity disorder. Transgender is a rare phenomenon in blind children and in children living in a jungle and not exposed to woman's high heels and makeup. Transgender is more common in sighted people who live in large cities full of colorful attraction and freedom of expression. Children brought up without fathers or without father bonding or a caring father who is capable of building up a close, trustful relationship with his child may become more attached to their mothers. Most infants have no sense of gender, and the way parents communicate with a child can condition their feminine or masculine behavior. If the mother is the main comfort for her boy, the boy may *feel* more comfortable to copy his mother's emotions and behavior. Similarly, if the parents kept treating a female infant as a male, she may grow up with a masculine behavior. Cross-gender behavior may increase during boredom, loneliness, or stress; and a male child may start to experiment with feminine behavior alone or with other female peers. After puberty, the surge of sex hormone may further change children's perception and belief system. Without support, men with gender identity disorder may develop a fixed belief that they were born as a female and/or feel of a woman being inside them. Gradually, they start to dislike their gender and/or start to perceive and react differently toward the same gender. In the liberal culture, they may behave and dress in the way that make them feel comfortable. As I mentioned, any behavior which leads to excitement or pleasure can lead to obsession and compulsion. Few adolescents may develop desire and curiosity to copy the pleasure they see in a woman's face when they wear glamorous clothes and makeup. Gradually, their curiosity changes to obsession and compulsion of wearing glamorous, extravagant sparkling female clothes, makeup, jewelleries, and hairstyles. Such behavior may give them internal feelings of pleasure and excitement and an escape from their boring or stressful reality. They may be conditioned to the extreme feminine behavior or may copy the behavior of some movie actresses. Furthermore, some cross-gender people develop a strong desire to change their gender by surgical operation and by hormonal therapy. This is more common in liberal cultures where people have freedom of expression. It is, however, a rare phenomenon and a stigma in the conservative religious culture or in the jungle.

## Paraphilia

Paraphilia is any sexual desire beyond normal cultural values. It may include abuse of living and nonliving objects without consent. However, cultural values are changing; and in the last few decades, some of paraphilic behavior has become acceptable in certain cultures such as exhibitionism, fetishism, frotteurism, and voyeurism. Paraphilia is more common in men than in women, and if it continues for few months, it may lead to obsession and compulsion and can affect people's personal and social lives. Extreme sexual behavior may remain a mystery. However, without law, religion, or moral conditioning, the human being is able to learn or copy any sexual behavior. Apart from rare congenital cases, paraphilia is a product of cultural conditioning (family, culture, or media). Spoiled children who are brought up by dysfunctional parents may grow up to be moody or a loner with radical ideas and behavior. At adolescence, some of them start to experiment with different sexual fantasies in order to fulfil their boredom and selfish needs. The pleasure of the fantasy can change into obsession and compulsion, which makes them addicted to one or more types of paraphilia. The prevalence of paraphilia is unknown as most of cases are illegal. However, behavior is contagious and conditioning. The incidence of paraphilia would increase in cultures who offer freedom or access to extreme sexual behavior at an early age. Paraphilia many include the following behavior:

1. Masochism or sexual arousal by receiving pain
2. Sadism or sexual arousal by inflecting pain
3. Bestiality or practising sex with animals
4. Necrophilia or practising sex with the dead
5. Pedophilia or sexually abusing children
6. Killing and/or eating the sexual partner

The last two examples are extreme. Certain men are conditioned with fixed beliefs that prostitutes or homosexuals are sinners, and they may kill them before or after having sex with them. Some of them may freeze and eat the body of their victims. However, pedophilia is the more common extreme paraphilic behavior. Most pedophiles are males, and they may also kill the child to cover their crime. Young children have immature sanity elements, and most of them do not understand criminal behavior or can't differentiate between sex and abuse. They may feel powerless in the hands of an abusive pedophile who spoils or grooms children. However, child abuse usually starts in the family. A dysfunctional family, or careless parents, may emotionally and/or physically abuse their children or leave them without care, love, education, or supervision. Such vulnerable children, with no supportive family, have no power of judgment, and they can't recognize right and wrong. Pedophiles usually groom vulnerable children by temptation, gifts, or force and/or threatening. Pedophilic behavior may start in the family, inside or outside schools, where physically powerful teens sexually abuse a vulnerable powerless boy for months or years by threats and abuse. Pedophilia is also common among religious men who are supposed to teach children moral values in the temple of God. The victims usually feel fear and shame to talk about their abuse to their parents, peers, teachers, or anyone. Their secrets may stay with them till death. They may live their lives envious of other normal or nonsexually abused children. The stigma provokes many negative emotions in the abused child and may affect their personal or social lives and their sexual orientation. Some victims of rape may commit suicide as they feel powerless to escape from their mental suffering.

# Impact of Sexuality

Sex can produce the human generation and can destroy them too. Life may not be excited without sex, but sex can cause the loss of millions of lives by STI epidemics. It can also create personal, social, ethical, medical, psychological, legal, economic, and political problems. Nonetheless, sexual lust or "testosterone urge" has pushed men to cross all cultural, religious, and legal barriers for temporary pleasure. In conservative culture, sex outside marriage may cause distrust, hate, anger, domestic violence, separation, and divorce. Also, causal sex without using a condom can cause HIV and its stigma, shame, depression, and fear of death. The stigma of HIV infection may last even after the death of the patients and may affect their families.

Many impacts of human sexuality have been highlighted in this book, but the impact of sexual lust on humanity has been alarming in the last twenty-five years as results of the AIDS pandemic. Most of these deaths might be avoided if men had worn condoms. However, despite the availability of the condom, some people are still not using it, and the prevalence of STI/HIV is still increasing. Apart from unsafe or unprotected sexual behavior, other factors have contributed to the spread of the HIV/STI such as illicit drugs, nitrite inhalants, excessive alcohol intake, accessibility of the sex venues and global communication, cellular phone, Internet chat rooms, affordable travelling facilities, sex trade through satellite channels, poverty, boredom, stresses, curiosity, greediness, political instability, poor public health service to control STI/HIV, wars, immigration, sex slavery, or sex trafficking and prostitution.

More than thirty bacterial, viral, and parasitic pathogens are transmissible sexually.

The common bacterial STIs:

- Chlamydia (*Chlamydia trachomatis*)
- Gonorrhea (*Neisseria gonorrhoeae*)
- Syphilis (*Treponema palladium*)

The common viral STIs:

- Genital warts, which are caused by human papillomaviruses (HPV)
- Genital herpes which is caused human herpes viruses (HSV)
- Viral hepatitis B
- HIV

The common parasitic STIs:

- Trichomoniasis (*Trichomonas vaginalis*)
- Pubic lice or crabs (*Phthiriasis*)
- Scabies (*Sarcoptes scabiei*)

Nearly a million people acquire STIs, including HIV, every day. The World Health Organization (WHO) estimates that every year, more than 340 million new cases of the curable STIs such as syphilis, gonorrhea, chlamydia, and trichomoniasis occur throughout the world in men and women aging from fifteen to forty-nine years old. Millions of viral STIs occur annually, attributable mainly to HIV; genital ulcers, mainly by human herpes viruses (HSV); genital warts, mainly by human papillomaviruses (HPV); and viral hepatitis. Globally, STIs including HIV constitute a huge economic burden, especially for developing countries where they account for 17 percent of economic losses caused by ill health. Infection with genital warts (human papillomavirus) is one of the causes for developing carcinoma of the cervix, which is killing around 240,000 women worldwide per year. Equally, hepatitis B viral (HBV) and hepatitis C viral (HCV) infections can cause cirrhosis and hepatic cancer. Untreated gonococcal and chlamydial infections in women may result in pelvic inflammatory disease (PID) in up to 40 percent of cases, and around 40 percent of ectopic pregnancies can be attributed to previous PID. One in four of these PID cases may cause infertility in women. While most of STIs are transmitted through sexual intercourse, transmission of STIs/HIV can also occur from mother to child during pregnancy (vertical transmission). Up to four thousand newborn babies become blind every year because of eye infections attributable to untreated maternal gonococcal and chlamydial infections. Untreated syphilis can result in neurological, cardiovascular, and bone diseases later in life. Untreated early syphilis during pregnancy is responsible for around 14 percent of neonatal deaths, and untreated genital herpes in pregnant women can cause a higher neonatal mortality rate during delivery.

The current global figure of people infected with HIV is more that forty million (around thirty-seven million adults, three million children). AIDS had killed more than twenty-five million people in the last twenty-five years or since it was first recognized in 1981, and the number is still growing. Such loss of human lives may have not been reported if men wore condoms. The sad irony is that the human loss is mainly caused by the sane human who able to use condom or practise safer sex.

The socioeconomic costs of STIs and their complications are substantial, ranking among the top ten reasons for health care visits in most developing countries and substantially drain both national health budgets and the household income. Care for the sequelae of STIs includes screening and treatment of cervical cancer, management of liver disease, investigation for infertility, PID, care for perinatal morbidity. Apart from the above costs, there are enormous personal, social, and psychological impacts of STIs/HIV morbidity and mortality on the patient and their families, including conflict between sexual partners and domestic violence. In addition, most STI/HIV cases affect young and highly productive people; hence, there is a large economic burden and loss of productivity to individuals and to their nations. The loss includes direct costs of medical and nonmedical care and indirect costs of time spent on sick leave.

Most people that acquire STIs, including HIV, are asymptomatic; and they can transmit their infections "silently" without their awareness. However, symptoms caused by STIs vary. For example, genital infection with *N. gonorrhoeae* or *Chlamydia* can cause pain on urination with green or cloudy urethral discharge in men and acute or chronic lower abdominal pain with cloudy vaginal discharge in women. Genital herpes infection is the commonest cause of genital ulcers worldwide and causes recurrent painful genital ulcers, which heal without scaring, while chancroid is a common cause of genital ulcers in the tropical countries and also causes painful ulcers, which can result in extensive tissue destruction if treatment is delayed.

Apart from using a condom, treating other STIs and genital ulcers can reduce the risk of HIV transmission, especially among populations who are most likely to have a high number of sex partners. Also, offering HIV treatment and a caesarean section to HIV-positive pregnant women can significantly reduce the rate of vertical transmission of the HIV. Equally, offering HIV treatment to people shortly after unsafe sexual intercourse with HIV-infected person or after an accidental injury by HIV-infected needle can prevent HIV transmission. The latter is called postexposure prophylaxis (PEP), which gives a higher success rate if it is offered within the first few hours of the exposure to the HIV. Nonetheless, the stigmatization associated with STIs/HIV is the main ongoing and powerful barrier to the implementation of STI prevention, especially in developing countries. STI patients may feel embarrassed to visit their local practitioners or hospital, fearing the breach of their confidentiality, or they may not be able to pay for sexual health checkup, treatment in the private sector, or the high cost of HIV therapy. Furthermore, people may have difficulty in notifying their sexual partners about their STI/HIV, fearing rejection, separation or violence; and their sexual partners may unknowingly spread the infection to other sexual partners, which may further impair the public health efforts to control STI/HIV (Ref: 6).

# Summary

Without sex, there would be no self; and without senses, there would be no feeling, pain, or fear. The human being is created with superior sanity elements that makes him the most intelligent and conditioned animal. The human child can be conditioned to be a creator or a criminal. Sanity and the senses, especially vision, are the main source of human civilization, but also the source for wars and conflicts. Human race has evolved from a family into tribes and cultures or nations. They introduce values and traditions, but never united in one nation. Their selfish instincts, belief systems, anger, and desires make them insecure and disunited. Equally, fear from failure, loss, stigma, law, sin, or abuse plays a major part in humans' secretive behavior and sexuality.

Most of human conflicts start with "sensory" abuse by disrespecting other people's senses "feeling." Sensory and/or power abuse can trigger many negative emotions such as distrust, hate, anger, and violence. Abuse of powerful people or group power is the most common source of *acceptable* human abuse. Equally, the media can condition people's belief systems and sexuality and make *unacceptable* behavior *acceptable*. Nonetheless, most child abuse starts by parents who fail to look after children's emotional, educational, and financial needs and/or teach their children to hate other races, faiths, or subfaiths. I hope that parents, schools, and the media can teach children to respect each other regardless their backgrounds and enable them to talk about their needs and/or abuse without fear. Although, humans have never been united, I believe that they can be united by respecting their senses "feeling" and their rights to survive in peace. I would love for you to join a "united humanity" network that communicates people all over the world in respect and peace to help every one living in our environment.

# References

1. Joined For Life—Conjoined Twins, Abby & Brittany Hensel turn 16. Added: September 13, 2007, http://video.google.co.uk/videoplay?docid =451087850499218280&q=conjoined+twins%2C+abby+%26+brittan y+hensel&total=4&start=0&num=10&so=0&type=search&plindex=1, accessed on 2/15/2008 part 2 of the sixty minutes interview of brothers. Cause of being gay. Added: August 08, 2007 http://video.google. co.uk/videoplay?docid=-85070799172647387 36&q=cause+of+being +gay&total=251&start=0&num=10&so=0&type=search&plindex=0, accessed on 2/15/2008

2. Ellis, Albert. *Anger, how to live with and without it.* Citadel press books, Kensington publishing corp., 2003.

3. It's an amazing toy for an Indian boy. A boy on a snake. Added: June 06, 2007. http://video.google.co.uk/videoplay?docid=-7066549387974517 648&q=boy+snake&total=644&start=0&num=10&so=0&type=search &plindex=0, accessed on 2/15/2008

4. Melzack. R, Wall. P. D, Ty. T. C, Acute pain in an emergency clinic: latency of onset and descriptor patterns related to different injuries. *Pain* 14 (1982): 33-43.

5. www.who.org

# Index

# V

# W

www.ingramcontent.com/pod-product-compliance
Lightning Source LLC
Chambersburg PA
CBHW022108170526
45157CB00004B/1534